Responding to "Routine" Emergencies

Responding to "Routine" Emergencies

Frank C. Montagna

> **Disclaimer**
>
> The recommendations, advice, descriptions, and the methods in this book are presented solely for educational purposes. The author and publisher assume no liability whatsoever for any loss or damage that results from the use of any of the material in this book. Use of the material in this book is solely at the risk of the user.

Copyright © 1999 by
PennWell Corporation
1421 South Sheridan Road
Tulsa, Oklahoma 74112-6600 USA

800.752.9764
+1.918.831.9421
sales@pennwell.com
www.FireEngineeringBooks.com
www.pennwellbooks.com
www.pennwell.com

Marketing Manager: Julie Simmons
National Account Executive: Francie Halcomb
Director: Mary McGee
Editor: James J. Bacon
Production/Operations Manager: Traci Huntsman
Cover Designer: Steve Hetzel
Book Designer: Daniel Krebs

Library of Congress Cataloging-in-Publication Data

Montagna, Frank C., 1949–
    Responding to "routine" emergencies / Frank C. Montagna
      p. cm.
    Includes index
    ISBN 978-0-912212-81-4
    1. Fire extinction.   2. Emergency management.   3. Home accidents.
  I. Title.
TH9146.M66   1999   99-33297
628.9'2—dc21   CIP

All rights reserved. No part of this book may be reproduced, stored in a retrieval system, or transcribed in any form or by any means, electronic or mechanical, including photocopying and recording, without the prior written permission of the publisher.

Printed in the United States of America

13 14 15 16 17   21 20 19 18 17

About the Author

Frank C. Montagna, a twenty-nine-year veteran of the Fire Department of New York, has served as an officer for the past twenty years and as a battalion chief for the past twelve. He has taught at the New York City Probationary Firefighters School and has been the Officer in Command of FDNY's Chauffeur Training School. He is currently the battalion commander of Battalion 58 in Brooklyn.

Chief Montagna holds a bachelor of fire science degree from John Jay College, where he has also taught firematics as an adjunct professor. As an author, his articles have encompassed many subject areas in the field of firefighting. He has been published in *WNYF* magazine and in *Fire Engineering* magazine, and he is a contributing editor for the latter. Trained by the International Association of Fire Chiefs as a CO response instructor, he has lectured extensively in the United States and Canada on the subject of carbon monoxide. Montagna holds seminars on firefighting-related topics and has spoken at the Fire Department Instructors Conference and the New York City Fire Department Institute's Structural Firefighting Seminars.

He resides on Long Island, New York, with his wife and daughter.

Dedication

This book is dedicated to Lieutenant Joseph P. Cavalieri and Firefighters James F. Bohan and Christopher M. Bopp. All three were working in Ladder Company 170 of the Fire Department of New York on December 18, 1998, when they made the supreme sacrifice. They were doing everything right when suddenly everything went wrong. They are remembered, and they are missed.

Acknowledgments

In the fire service, knowledge is spread around and shared. Whenever a postfire debriefing is held, we learn what each firefighter saw, thought, and did at a particular incident. We learn what mistakes were made and what was done correctly, and we discuss how the operation could have been improved.

When we attend a drill or firefighting seminar, we learn from the experience of others. At social gatherings, firefighters are notorious for forming groups and talking shop. They may discuss their latest fire, a civilian death, a firefighter's injury, a new tactic that either worked or failed. The knowledge gained in debriefings, drills, seminars, and informal conversation is stored away in each firefighter's brain, because each one of them knows that, to stay alive in a burning building, he needs knowledge of fire behavior, building construction, hazardous materials, operational procedures, and many other related and sometimes obscure bits of information. Most of all, a firefighter needs experience to make the life-and-death decisions required to successfully navigate a fire scene. Unfortunately, each of us can only be present at so many incidents. Thus, we can only gain limited personal experience.

New firefighters naturally have less experience than their older counterparts, but they respond right alongside their more experienced brethren whenever an alarm sounds. The probie is at a distinct disadvantage because of his lack of knowledge. As a result, he often pays close attention to the stories told 'round a keg of beer by the old timer.

Throughout his career, this fledgling firefighter will hear and tell many more stories, until he is the old timer himself, recounting his past exploits. His knowledge will increase far beyond the level of his direct experiences alone, and he in turn will increase the knowledge of others. Like all firefighters, I have listened to stories and asked questions throughout my career. I have also told stories and answered questions. I have learned much. It is my hope that, with this book, I can share some of that knowledge with others.

I want to thank all of the firefighters with whom I have worked for freely sharing their knowledge with me. Without them, I might not be here today, much less have been able to write this book. While I have researched the topics presented here, much of the information has been garnered from firefighters who gladly shared with me their expertise on specific topics. Firefighters and other professionals have read my manuscript throughout each stage of its development and have given me insightful suggestions as to how to improve it.

Although there are too many deserving of thanks to be mentioned here, I must give special recognition to the generous people who took considerable time to review, comment on, or provide information for various portions of this manuscript. They include John Campana of Brooklyn Union Gas; Bob Damato, Kevin Rogers, and others of Consolidated Edison; Frank Capoza for reviewing this manuscript and offering suggestions and encouragement; and Pat Coughlin of the International Association of Fire Chiefs (IAFC) for allowing me to participate in the IAFC Carbon Monoxide Seminar.

Ted Goldfarb got me interested in lecturing on fire-related topics, which eventually led me to writing articles for *Fire Engineering* magazine. Larry Petrillo taught me how to create and deliver a presentation. Bill Manning of *Fire Engineering* encouraged me to write this book. Due acknowledgment goes to all of them.

A special thanks is in order for my wife, Dessie, and my daughter, Katrina, who put up with my long hours at the computer. Thanks also go to my father, Cottello, a retired firefighter, who guided me into the fire service, and to my mother, Helen, who spent many hours making sure that I learned what my teachers had tried to teach me.

Contents

Introduction 1

PART ONE
COMMON CALLS 3

Chapter One
    Electrical Emergencies 5

Chapter Two
    Home Heating Emergencies 43

Chapter Three
    Natural Gas Fires and Emergencies 67

Chapter Four
    Water Leaks 93

Chapter Five
    Vehicle Fires 107

Chapter Six
    Kitchen Fires 127

Chapter Seven
    Mattress Fires 141

Chapter Eight
    Trash Fires 149

PART TWO
CARBON MONOXIDE 165

Chapter Nine
    The New Response 167

Chapter Ten
    The Medical Aspects of Carbon Monoxide 175

Chapter Eleven
　　The Carbon Monoxide Emergency............ 183

Chapter Twelve
　　Home CO Detectors 209

Chapter Thirteen
　　UL 2034.................................. 223

Answers to Study Questions..................... 237

Index.. 241

Introduction

As firefighters, we repeatedly respond to certain types of incidents. The routine calls aren't the most exciting runs that we go on, nor are they the kind that we sit around and discuss at social gatherings and in bull sessions. Sometimes we don't even give much thought as to how to handle them or to the threat that they pose. Much has been written about structural collapse, backdraft, confined spaces, and hazardous materials, but where is it that we spend most of our time? The main portion of a firefighter's professional life is spent putting out car fires, removing burning pots from stoves, and shutting down malfunctioning oil burners.

A simple mattress fire is nothing compared with a multiple alarm in a paint factory that results in structural collapse and a river of burning paint flowing down the street, but which scenario will we experience more frequently? Most of us will never even see a burning paint factory, but mattress fires occur in every community, sometimes with deadly results. You've probably received training in building collapse and exposure protection, but have you trained recently on mattress fires? Preparing for the major events is essential, of course, but we must also take the time to analyze and train on those calls that we respond to over and over, day after day. How much water should you use at a mattress fire? Should you leave the mattress in the building or remove it to the street? Should you take it down the stairs or throw it out the window? Have you ever even thought about what you do at a mattress fire and why you do it?

As a young firefighter, I regularly responded to several car fires a tour, as well as to an equal number of food-on-the-stove incidents. I rarely gave a thought to the dangers posed by these and other routine responses. I can't tell you how many electrical emergencies and gas leaks that I have confronted that were truly minor incidents, but every so often, one would turn into a major incident with the potential for doing structural damage and even causing injury or death. The question is whether we are alert to the potentialities of such a routine response or whether we remain complacent until the situation gets out of control and overtakes us.

As my time in the department grew longer and as promotions increased my responsibilities, I became more and more aware of these hazards and potentialities. Gradually I developed a more cautious approach to such incidents. The young firefighters that I now supervise, as well as some young officers, exhibit the same cavalier attitude toward routine responses that I once did. Undoubtedly they consider my caution to be an annoyance. I understand such an attitude, for I once felt the same way. One chief with whom I worked early in my career routinely made us stand by for what seemed an unreasonable length of time at odor-of-smoke incidents, until either the truck company found the source or he was satisfied that there was no threat. Often that meant missing a meal or a TV show or, worse, missing sleep. I couldn't understand why, if there was no fire, we had to stand fast in the street. Why not just send us home and, if absolutely necessary, have one company search for the source of the odor? Now, as a chief, I am undoubtedly just as annoying to the firefighters who work with me. Although most of these smoke-odor calls turn out to be nothing, I have, after a prolonged search for an odor, discovered potentially dangerous situations that might otherwise have resulted in hazards to the occupants. Now when I am tempted to give up on the source of an odor in a home, I look at the occupants and think what the consequences might be if a fire breaks out after we leave. What if they have gone back to bed and are asleep at the time? What if there is loss of life? Could I live with that? Could I explain it in court? In March 1995, *Fire Engineering* magazine published my article "Odor of Smoke." Shortly thereafter, I was asked to testify (I declined) against a fire department that had responded several times to an odor of smoke in a commercial occupancy. Each time they responded, they found no cause for the odor. Sometime later that night, after the occupants had closed up and gone home, a fire broke out that gutted the structure. The cause was determined to be electrical, and it was believed that the same electrical problem had caused the odors detected earlier in the day. Luckily, this was a commercial building and not an apartment house full of sleeping residents. No one was killed, but the chief and the department had to answer in court for their actions and inactions.

I wrote this book to heighten firefighter awareness of these routine incidents, to point out that a hazard potential exists at them, and to open a discussion on the standard tactics used for them. It doesn't matter whether yours is a paid or a volunteer department, nor does it matter whether your department is large or small. The potential hazards will be present in any case.

The topics discussed in this book aren't as exciting as backdraft, collapse, hazardous materials, or a host of other subject areas. Still, you'll find yourself at more food-on-the-stove incidents than backdrafts, and at every food, gas leak, or heating response, you'll be put at risk of injury. Doesn't it make sense to spend some time discussing and training for what you do most?

In truth, your "routine" response is not routine. No response is. Every type of response should be discussed and dissected, and you should develop appropriate plans for them. This way, you can minimize the dangers to firefighters and the public alike as you effectively, professionally perform your job.

Part One
Common Calls

Chapter One
Electrical Emergencies

Firefighters are called to homes for a variety of electrical problems, whether flickering lights, electrical odors, or to extinguish fires of electrical origin. The cause of the problem can lie in the building's wiring; it can be the result of a faulty appliance or one that is being misused, or it can be due to some external force such as lightning, mechanical damage, or a water leak. Typically, at nonfire emergencies, you will be met with an electrical odor when you enter the premises. The odor may or may not be accompanied by haze or smoke, but in either case, the source of the odor probably won't be obvious. Now is the time to interview the caller to determine what was happening in the house prior to the development of the odor.

Regardless of the cause, you must determine what the problem is and locate the portion of the electrical system that is affected. Once the problem has been identified, you must remove the hazard. This can mean shutting down the power to an entire building, an apartment, a line of apartments, a room, or it can simply mean pulling the plug of the offending appliance. You shouldn't fix the problem; rather, you should leave the task of restoring power to licensed electricians. Even if one of the firefighters happens to be an electrician, you should still only mitigate the hazard, not make repairs. In some municipalities, firefighters issue orders to have a defective electrical system tested by a licensed electrician before restoring power. This is a good idea. In any case, if you shut down the electrical supply to a house or appliance, you should warn the occupant not to turn it back on until it has been checked and deemed safe.

ELECTRICAL ODORS

Calls for unusual odors in residences and commercial occupancies are common. When you arrive at such a call, you may smell the odor outright. Electrical odors are easily recognized by firefighters who have experienced them. Otherwise, the odor may have dissipated prior to your arrival. You can inadvertently cause the odor to dissipate by repeatedly opening the door, allowing countless firefighters to enter the building. Sometimes the smell from your own smoky turnout gear can interfere with your ability to recognize a reported electrical odor.

An odor that is evident when you enter an occupancy can become undetectable as your olfactory system becomes desensitized to it. If this occurs, you might have to bring in a fresh nose from the street. For this reason, it is a good idea to keep several firefighters outside of the building. In fact, it doesn't make much sense to have an entire first-alarm assignment traipse into a private dwelling to search for an odor of smoke. Only a few firefighters are needed. The rest can stand by outside as potential reinforcements. On the other hand, a large commercial structure might require the commitment of all of your available personnel to track down the source of the odor.

If you find a haze of smoke with a strong electrical odor, consider using SCBA. There have been instances where exposure to such a haze has resulted in firefighters experiencing breathing difficulties at a later time. If the insulation of the burning wire contains PVC, it will give off hydrogen chloride gas. If the call is in a commercial occupancy, a telephone exchange, or a building that houses a transformer, the smoke can be quite dangerous. Polychlorinated biphenyls, contained in some transformer coolant oils, can give off carcinogens when they burn.

ELECTRICAL LIGHTING EMERGENCIES

Fluorescent Lights

Overhead fluorescent lights are a common source of electrical odors. Typically, they will result in a call for an electrical odor or haze of smoke. When you arrive, you might recognize the odor as that of an overheated ballast. If so, then the rest is easy. Look up at the ceiling for a fluorescent fixture that is blinking, dim, or not working at all. Question the occupant and find out whether he has shut off any of the lights. Ask whether any of them were flickering and for how long. Also ask whether any have recently stopped working.

Fluorescent lights require ballast to function. A typical ballast transformer consists of fine copper wire wound around an iron core. This is coupled to several electronic components, including a capacitor—normally a small, cylindrical aluminum device designed to store and release electricity. These parts are embedded in pitch to hold them in place, drown out electrical hum, and lower their operating temperatures, thus increasing their life span.

All fluorescent lights employ some form of ballast device, although not all contain pitch. Most ballast devices are separate components, but in some compact lamps, the ballast is an integral part of the lamp. The type typically found at an electrical odor response is not integral, but rather, a separate component. The ballast can often be found encased in a metal box attached to the reflector hood. This box may be visible once you remove the plastic lens from the fixture. Otherwise, you may have to remove the fluorescent tubes and then a metal panel. The ballast can overheat because of a short in the transformer core or failure of the capacitor, the light starter, or one of the fluorescent lamps. As a result of repeated attempts to start the light, the ballast heats up. This causes the surrounding pitch to heat up, giving off the distinctive odor of the overheated ballast. If the heating progresses far enough, smoke is generated. As mentioned above, flickering, dim, or nonfunctioning lamps can indicate that the ballast is overstressed and heating up.

A positive check is to feel the ballast. Normally, ballast devices run quite hot, about 140°F. If one is so hot that you can't keep your hand in contact with it, then it is likely to be the culprit. Use caution when touching them so as to avoid burns. Also, be careful as you remove the cover from the ballast. If you move the fixture, melted pitch can drip down on you. Dripping pitch is in itself a good indicator that a particular fixture is the source of the odor. It is possible, in ballast devices manufactured before 1979, that the capacitor contains a small amount of PCB-contaminated dielectric fluid. If the fixture heats to the point that the capacitor ruptures, the dielectric fluid can leak out, and if the fluid is burned, the PCBs will produce dioxin-contaminated smoke. After the ballast device cools, the melted pitch will again harden. If you still see leaking fluid, possibly what you are seeing is the leaking dielectric fluid. Take precautions that you're neither burned by the melted pitch nor contaminated by contact with the fluid. Also be careful not to inhale the smoke generated by the ballast. Although the small amount of PCBs in ballast isn't thought to be dangerous, prudence dictates the use of SCBA. New fluorescent lamps, those manufactured after 1968, should contain thermal protection that will prevent the ballast from overheating. As the older lights are replaced, the number of such calls should decrease. Of course, thermal protection can fail and result in an overheating ballast.

If you have found the source of the odor to be an overheating ballast, remember to check the ceiling space above the fixture. It may be possible, over time, for the ballast to become hot enough to ignite combustibles that are in contact with the metal mounts. If the light is part of a suspended ceiling, popping a tile from the ceiling is sufficient. Be especially cautious if the light is in contact with fiberboard ceiling tiles. Often these tiles are combustible and can be ignited by an overheated ballast. If the fixture is mounted on plasterboard or a plaster ceiling, you may need to poke an examination hole in the ceiling or even remove the fixture so as to visually inspect the area. The presence of any intact plaster-type material should be enough to protect combustibles in the space above the ceiling, making ignition unlikely. Still, if there are holes in the plasterboard behind the light, and if some other electrical problem such as arcing is involved, ignition is a possibility.

Once you find the faulty fixture, you must mitigate the problem. Removing an eight-foot fluorescent lamp from its fixture should halt the flow of electricity and thus the overheating of the ballast. If the tube is later replaced, however, the problem will reoccur. Removing a fluorescent lamp smaller than eight feet will not stop the flow of electricity. To be sure the fixture is safe, disconnect the wires supplying electricity from all of the overheating ballast devices that you encounter, regardless of the length of the tubes. This way, you can be assured that the overheating will stop; the hazard will be abated; and that the occupant won't just replace the tube, causing the overheating to resume. Remember to shut off the power supply to the lamp before you work on it and to tape or place wire nuts on wires that have been severed so as to leave the fixture in a safe condition for the residents.

Often the source of the electrical odor won't be obvious. At one such incident, we encountered an odor of smoke in an apartment above a grocery store. It smelled like ballast, but the apartment contained no fluorescent lights. A firefighter checking the store below the apartment found no odor but noticed a flickering fluorescent lamp. He raised an A-frame ladder to the fixture and checked the ballast, which was extremely hot. There were several holes in the plaster ceiling above the fixture, through which the odor had been sucked out of the store and into the apartment above.

An overheating fluorescent fixture that isn't disconnected will continue to overheat. It can overheat to the point that the pitch contained in the ballast melts, ignites, and drips flame down onto any combustibles below. The vast majority of calls involving malfunctioning fluorescent lights, however, do not involve fire. Usually it is a simple matter of locating the offending fixture and disconnecting it. However, the potential does exist for fire and for smoke injury, and in some cases, the simple odor of smoke can result in a haz mat incident. As in much of what we do, it is better to be overcautious than to take such calls for granted.

Recessed Lighting

Many modern homes safely utilize recessed lighting. Occasionally, as a result of a fault in the light itself or perhaps improper installation, an odor or fire develops and the fire department is called. Older recessed lights aren't thermally protected—Underwriters Laboratories didn't require thermal protection for approved recessed lighting until 1982. Without proper thermal protection, a recessed incandescent light can cause an odor of smoke. Even one that has been installed properly can cause smoke or fire if its high-limit switch fails. Again, you should always interview the occupants to find out what was going on in the house prior to the onset of the odor.

A recessed incandescent light that had been switching on and off every ten minutes or so could indicate a thermally protected light that was overheating and shutting down as a result. Each time such a fixture cools off, it can switch itself on again. Ask the occupants whether any such lights had been operating intermittently before they placed the call.

If the thermal protection failed, the fixture could overheat, resulting in smoke or fire from nearby insulation and wood framing. Even a properly functioning recessed light can cause a fire if it is installed too close to combustibles or if insulation is installed over it.

Lamps

A plastic bag placed over the top of an incandescent lamp shade can heat up and ignite. The occupant, seeing smoke or smelling an odor, will call the fire department. By the time the firefighters arrive, the plastic bag might be gone, leaving smoke in the room and only residue on the shade or the floor to indicate the source. Check for residue or ashes when faced with a baffling odor or smoke condition.

Halogen Lamps

Halogen lamps, popular for their bright light, produce high temperatures as well. Recently, a multiple-alarm fire in a Manhattan luxury high-rise was the result of a halogen lamp that inadvertently tipped over. The bulbs in these lamps can reach temperatures in excess of 1,000°F. Even if the lamp doesn't tip over, the bulb can rupture if it hasn't been installed properly, dropping the hot filament onto combustibles below, igniting them. Manufacturers of these lamps are offering to retrofit them with protective guards intended to reduce the fire danger.

ELECTRICAL APPLIANCES

What can go wrong with an electrical appliance? If it has a motor, the motor can heat up because the airflow to it is blocked, because it is overloaded, or just because it has reached the end of its life. If it has a drive belt, the belt can slip, causing friction, which in turn causes heat, or possibly smoke or an odor. A slipping or broken belt can be the reason that a fan isn't delivering cooling air to an overheating motor. An appliance transformer can malfunction and deliver excess voltage to the appliance, causing it to heat up. Many appliances have high-limit switches that should turn them off before the appliance can generate enough heat to create a fire hazard, but sometimes they malfunction. Small appliances like toasters and hair dryers have thermal switches that cut off the current if they overheat. If the high-limit or thermal cutoff switch malfunctions, the appliance can heat up enough to cause an odor, if not smoke and fire. Again, this is where your interview can direct you to the cause of the odor.

Fire Department Response

Ask What Appliances Were in Use at the Time and Check Each One. Even check appliances that the resident says weren't in use. Ask whether any appliance stopped working.

Check for an Odor. Is there an odor near an appliance? A defective appliance may give off an electrical odor that permeates the room or even the building, but the odor should be strongest near the device itself.

Check for Heat. If an appliance is hot or if there is an odor, check the on/off switch. If it is on and the appliance isn't functioning, check the plug. It the appliance is plugged in, check the fuse or circuit breaker. Has it blown or been tripped?

At a call to a large postal facility, firefighters were baffled as to the cause of an electrical odor. They checked outlets, air conditioners, heating units, and lighting fixtures, all to no avail. The source of the odor turned out to be a malfunctioning video display terminal. A fire officer noticed that one terminal screen was black and that all of the others were illuminated. Moreover, it was plugged in, it was turned on, and it felt hot to the touch.

If you suspect the appliance, unplug it and examine it for signs of fire or smoldering wire insulation. Is some combustible material in contact with hot wires? You don't want to walk away and leave a smoldering appliance behind. If you find fire in the appliance, unplug the device and extinguish the fire. Often a short burst from the water or dry-chemical extinguisher is all that's needed. Check for possible extension. For an overheating motor, all that may be needed is to cool the motor. This can be done with a water extinguisher or wet towels. If the power is removed from the motor and it isn't in contact with combustibles, it can cool down unaided with no threat of fire. If flame is found, check the nearby area for extension.

Consider the Unusual. At a call to a home for a report of an odor of electrical smoke, a crew of firefighters and I conducted a thorough search and found nothing. Initially we had detected an electrical odor, but it dissipated. Just before giving up, we noticed a new odor. It was an odor of smoke, but this time it was not electrical, and a haze was now developing in the kitchen. A new search quickly located the source. Accumulated dust and debris behind the refrigerator that was clogging the motor and coils had been ignited by the initial electrical problem. By the time we moved the refrigerator away from the wall, we had active flames, which we quickly extinguished with a water extinguisher.

One common source of electrical odor isn't electrical at all. It occurs when the drive belt of an appliance overheats. The smell of heating or burning rubber mimics that of an electrical problem. Check the belts on appliances, if so equipped. This includes items such as the dishwasher, clothes washer, and dryer. If the machine is overloaded or if the belt is worn, the belt may slip, causing smoke from the friction.

Be careful if you plan to move a washing machine full of water. At eight and a half pounds per gallon, a twenty-gallon tub of water weighs 170 pounds. Add to that the weight of the machine. If there is an odor only with no smoke, unplugging the washer should be enough. Wait a few minutes to see whether the odor dissipates. If it gets stronger or if smoke appears, then you must look further. Consider emptying the washer so that you can easily move it and make a thorough investigation. In one instance, when we moved the washer, we found the powdered remains of the drive belt on the floor. The belt had gotten so hot that it burned, leaving only a powdery residue behind. Also consider the possibility that the washer isn't the source of the odor. Don't end your investigation just because you suspect an appliance of being the cause. What if you're wrong? What if more than one appliance has malfunctioned because of a problem with the home's electrical system? Make a thorough search of the area for other possible sources.

Clothes dryers can often be problematical. The source of the odor can be from an electrical fault, an overheated belt, lint ignited in the exhaust ductwork or a clogged lint collector, or clothing dried too long or at too high a temperature. Sneakers and other rubber items will give off an odor before they actually ignite. Don't forget to look on the inside of the drum for patches of melted rubber or synthetic clothing that may have been the source of the odor.

The strangest clothes dryer incident that I ever responded to was in a high-rise apartment complex. We responded to a call for smoke emanating from a manhole that was a hundred feet behind one of the buildings. The cause was apparently rubbish below a sewer grate. When we directed a hose stream into the hole, we noticed smoke puffing out of the first-floor laundry room window of a nearby building. It turned out that the manhole was actually a vent for the clothes dryers. The air from the dryers was piped into a small room adjacent to the laundry room. From there, it was sucked by a large fan into an underground

The interior of this commercial dryer drum shows scorch marks and ashes.

Check the area below the drum of a commercial dryer for smoldering lint.

Check the entire run of the ductwork for extension.

passageway that led to the manhole, where it was vented to the outside. Lint had accumulated and ignited in the underground portion of the duct. To extinguish the fire, we had to turn off the vent fan and snake into the duct a charged 1³/₄-inch hoseline, with the nozzle open, from within the building while we simultaneously applied water through the manhole vent.

The accumulation of lint is a problem in home dryers as well as in commercial units. Once ignited, the lint will create a smoke condition. For a home unit, remove the lint screen. Check the interior of the machine and the vent duct for smoldering lint, and also check the area around the dryer and along the run of the ductwork for extension. A commercial laundry requires the same attention. Open the front panel of the dryer to look for smoldering lint, and check the vent duct. You'll have to check the entire run of the duct for smoldering lint, as well as extension. It isn't uncommon for the duct, heated by burning lint, to ignite structural wood, resulting in a full-fledged structure fire. A thorough check is required. A thermal imaging camera is useful for locating hot spots hidden in the duct or behind the ceiling or wall.

SPARKING OUTLET OR FIXTURE

Frequently firefighters are called to a home because of a sparking or smoking outlet or fixture. When you arrive on the scene, the occupant will typically point at it and describe how he saw sparks jumping out. If it is an outlet, it may show some discoloration, either burn marks or smoke stains on the cover plate. If it is a fixture, the discoloration may be around the mounting plate. In most cases, there won't be any fire, but you should take nothing for granted. Each such call requires a thorough examination.

As you enter the building, your nose will supply you with information. You can expect to smell an electrical odor, but if you detect the smell of burning wood, you may have more than an electrical emergency. You may, in fact, have a structure fire.

Begin your investigation by finding the occupant who first noticed the problem. This way, you'll get firsthand information, not secondhand misinformation. Have that person direct you to the suspected outlet or fixture. Ask what he saw. Did he see smoke, sparks, or flame? Inquire as to what was going on in the house at the time. What appliances were being used? What was plugged into the outlet? Did the lights in the house blink? Was any other appliance, fixture, or outlet affected? If so, the scope of the problem may include more than one outlet.

Once you have located the offending outlet or fixture, cautiously feel it and the surrounding area. Is it hot or warm to the touch? If no heat is found, simply removing the cover or mounting plate may suffice. The discovery of burned or melted wiring within the electric box indicates the need for further examination. Feel all around the general area. If you feel heat, immediately open up the wallboard and look for fire, char, or smoke stains. Usually a wall that is too hot to

14 • Responding to "Routine" Emergencies

This wall has been overhauled to allow for an examination of the studs. If the BX cable has been heated, the firefighter should follow its run, checking for extension.

hold your hand on indicates fire on the other side. If the wall near the outlet or fixture is warm to the touch but not hot, it may indicate overheating wires, an incipient fire, or a small fire that has self-extinguished or is just smoldering. In any case, make an examination hole just to be safe. The presence of smoke, fire, or char within the wall means that further examination is necessary. Consider, however, that the char or smoke stain may be from a previous fire.

Leave the wiring in a safe condition by taping or capping leads that have been exposed by your examination. Tripping the breaker or pulling the fuse that controls the fixture or outlet will stop the flow of electricity, thus removing the source of ignition and making the wires safe to handle. If the problem is widespread, you may have to kill the power to the entire apartment or building. Remember never to restore power once you have removed it. Restoring power is the responsibility of an electrician after he thoroughly checks the circuit for safety.

CAUSES OF ELECTRICAL FIRES

A number of electrical problems can cause fire to break out inside of a wall. Fire can start as a result of a damaged outlet, a loose connector, or worn or damaged insulation that has allowed the copper conductor to contact the metal of the box or the metal cable sheathing. This, combined with a poor or missing ground, can cause the cable to heat. It can heat suddenly, becoming cherry red, or it can heat over a long period of time, slowly charring wood with which it has contact. This charred wood can, after prolonged heating, burst into flame.

A wire can come loose in a junction box, causing a ground fault. It can charge the box or even the entire run of the metallic cable. It might overheat at the point of fault, or it might not. The cable can carry current to the next box and cause no problem there. The cable, still charged, can then carry current along the line to a box, where a loose connection causes it to heat sufficiently to ignite wood, insulation, or paper stuffed in the wall. What you now have is a fire remote from the original cause and possibly remote from the original fire area.

At one response to a report of hot walls, we found a fire in one bay, ignited by red-hot BX cable. We checked above and below the point of origin, as well as several bays to either side, but found no extension. We pulled the fuse that was affecting the involved wiring and then extinguished the fire. As we were taking up, however, a firefighter noticed smoke increasing on the floor above the fire floor. Obviously the incident wasn't over, as we had thought. Further investigation revealed that the electric fault had started a fire in one bay and then, skipping six bays, ignited another fire. Although we had checked for extension, we obviously hadn't checked far enough. How can you avoid making this mistake at electrical fires to which you respond?

Examining for Extension

Obviously, you can't open every wall and examine every wire in the building every time you're called to a home for an electrical problem. This is neither practical nor necessary. You must, however, bear in mind that fire can ignite remote from the area that you determine to be the point of origin.

Kill the Power. Once you have determined the cause of the fire to be electrical, pull the fuse or open the appropriate circuit breaker. This will do two things. First, it will remove the source of heating from the cable. Second, it will remove the danger of electrocution. If the armored metal cable is charged, anyone touching it can receive a shock. If a firefighter is standing in a puddle of water, not uncommon in our line of work, he could be electrocuted. Removing the power will prevent electrocution. At one incident, a red-hot cable ignited a stud that it passed through. A firefighter searching for additional extension six feet from the initial ignition point was burned by the heated cable when he touched it with the back of his ungloved hand. Killing the power will stop the heating of the cable, but it won't remove the heat that has already accumulated. To avoid painful burns, use caution when searching for heat in a cable.

Trace the Wires. Try to determine where the involved wiring is coming from and where it is going. Check the wall and ceiling along this run for heat, and check any outlets, wall, or ceiling fixtures that you encounter for heat or smoke. Pay special attention to where the wiring lies across or passes through beams or studs.

Open an Examination Hole When a Surface Is Hot. Use your sense of smell to determine whether wire or wood is burning. If the fire has entered any void, thoroughly check the void as a possible avenue of extension, as you would in any fire. An electrical fault can ignite a stud, and fire can travel up it to the cockloft in a balloon-frame building, resulting in a cockloft fire. Often, wires run into the attic or cockloft space and branch out into various rooms. Taking a look into these spaces can save you the embarrassment of returning to the scene at a later time for an attic or cockloft fire.

Don't Forget That Wires Travel Down as Well as Up. Fire can extend to or originate below what you perceive to be the original involved area, and burning embers can drop down a void, igniting combustibles below the point of origin. Wires can also pass through walls that are common to more than one apartment. Check both sides of the wall for fire extension. Although the fire might not have extended into the room or apartment that you're in, it may have ignited combustibles on the other side of the wall in the adjoining apartment.

Use a Thermal Imaging Camera. Wouldn't it be nice if you could see into the wall and locate any hot wires, smoldering beams, or active flaming? If you

A thermal imaging camera can detect fire and even hot wires behind a wall or ceiling. It provides an easy way of locating an overheating ballast in a large commercial occupancy. (Official FDNY photo.)

had this capability, you'd never miss any hidden fire. If you have access to a thermal imaging camera, you do have this capability. Scanning the walls with the camera will reveal any hidden pockets of fire. It can even locate overheated wires and smoldering beams and studs. I have successfully used this device to check for fire behind a tile wall in a school auditorium in which an electrical panel had burst into flames. It allowed me to depart the scene, secure in the knowledge that I hadn't left behind a smoldering time bomb. At the same time, it saved the expensive tile wall from needless damage.

ELECTRICAL SERVICE PROBLEMS

If the power company is unable to meet the demand for electricity, it sometimes reduces the voltage that it supplies to its customers. This can cause the service box and wires to heat up as the operating appliances demand more voltage than is available. The fire department can be called in such cases as a result of noise from the box, electrical odors, smoke, or a hot box or meter.

A fire department was called to a school for a report of an electrical odor. An investigation revealed that a pull box, a sealed electrical box located where the underground electric service entered the cellar of the building, was hot and smoking. This box was secured by a utility company lock. Fire department units on the scene removed the lock and opened the box to see what the problem was. The box contained several splices that connected the building's wiring to the supply wiring coming underground into the building. The firefighters correctly didn't disturb anything inside the box; rather, they stood by and periodically monitored it while waiting for the utility company to respond. The box, however, was left open, and later, the school's custodian went to close it. As he did so, the cover contacted a wire exposed as a result of the melted insulation, and it delivered a shock, the force of which knocked him down. Luckily, he wasn't seriously injured. Electric boxes locked by the utility company usually contain high voltages and may contain splices.

Responding to Electric Utility Incidents

The Fire Department Should Not Open Sealed Electrical Boxes. Even if smoke is emanating from the box, or if it is hot, you should not open such boxes. What you should do instead is find the main shutoff for the building and open it. This separates the incoming power source from the house wiring. The demand will end, and that should be enough to stop the overheating of the wires.

Open the Main Switch. If the incident is in a residential building, you can, if one is present, open the main switch, thereby cutting off power to the home. This should cut the power to the service box and allow it to cool down. It is, however, possible for power to be pirated into the building, bypassing both the meter and the main shutoff. Such a condition could leave some live wires in the home even after you shut down the main.

If the incident is in a commercial building using high voltages, opening the main switch may not be possible or advisable. Buildings that house heavy users of electricity and are supplied with 460-volt services will have the electric service in a locked room with a posted warning sign. These rooms contain very dangerous voltage. In New York City, such rooms require a sign saying *460V Keep Out. In case of emergency, call this number.* In no case should fire department personnel enter such a room. The utility company lock on the door should not be forced by firefighters. Inside, there is exposed wire and bus, and any contact with the charged elements in this room will cause death. In fact, it will not only kill you, it will continue to burn you long after you die. It will incinerate you. There is no safety device that will trip and cut off the power if you contact it. The utility personnel fondly say that it won't even know that you're there.

Usually there will be automatic fire alarms in these rooms that have a direct connection to the fire department. Recently, one of these alarms went off in a large municipal building. The fire department was called to the incident and proceeded to force the door. When the utility company emergency response crew arrived, there were ten firefighters milling around the room, oblivious to the danger. The alarm had tripped because of a dust condition created by a manhole explosion in the street. There was no need for firefighters to be at risk in the room, since there was no emergency. Even if there had been a fire in the room, no water should have been applied because of the danger, nor should any firefighter have entered, especially with metal tools. In such a case, the best course of action is to wait for the utility personnel to kill the power before taking action. Control any extending fire, evacuate the area, and ventilate, but do not enter the room or apply water to it.

Avoid Cutting the Wires. Cutting the wires should not be attempted unless absolutely necessary, and not without the proper equipment and training. Wires that supply electricity to the building carry high voltages. The wire cutters and lineman's gloves carried by many fire companies may not protect the firefighter against a deadly shock if they haven't been properly tested or stored, if they are damaged, or if they are used improperly.

If the overheating box is located before the shutoff, or if there is no shutoff and the wires are overhead, cutting the overhead wires will disconnect the building from its power source. Again, you must consider the possibility that electricity is being pirated into the building. Some large commercial buildings may be supplied with electricity at more than one point, and cutting one set of wires may not curtail the power. A safer solution than cutting the wires would be to monitor the area until the power company arrives. If fire is extending from the box, use a fog stream to contain it, but do not apply water directly onto or into the electric box. Monitor for carbon monoxide as well as fire extension, ventilate if necessary, and control any extending fire.

The heat from the burning and arcing wires in this meter box cracked the meter above. The fire was brought down using a dry-chemical extinguisher.

Remove the Meter Only if You Are Trained to Do So. Removing the meter from a building is best left to utility company personnel. They are trained and equipped to do so. Their protective equipment is regularly tested and is reliable. Typically, a fire company's lineman's gloves sit unused on the engine or truck for years. When the need for them arises, are they still serviceable? Have they been regularly tested? Have the firefighters been routinely trained in their use? A pinhole in the gloves could result in death.

If you plan to remove the meter and are trained and equipped to do so, first remove the electrical load from it, if possible, by shutting the home's main breaker on the service panel. Removing the meter should separate the structure from the source of electricity and stop any overheating of the wires. Remember, there may be more than one source of electricity and more than one meter supplying the building, so all electrical equipment and wiring must still be considered live. Another thing to consider when contemplating removing the meter is that removing it may *not* cut off the flow of electricity into the building. Recently I read a notice from a local utility company, warning of a new type of meter being used on residential buildings. The intended purpose of the meter was to allow the utility company to change the building's meter without interrupting the flow of current to the building. Removing this meter wouldn't result in interruption of electrical service unless specific additional steps were taken afterward. The same problem may exist at commercial buildings as well as residential ones. Do you have this type of meter in your district? Now would be a good time to check for it.

ENERGIZED ALUMINUM SIDING

One afternoon, we were called to a private home by an occupant who complained that he received a shock every time he touched the aluminum siding. Our investigation revealed that one of the overhead wires delivering current to the home had been damaged and was contacting the aluminum rain gutter. This charged the aluminum siding. When the homeowner touched the siding, he gave the electricity a path to ground and so received a shock. We removed everyone from the house and awaited the utility company's response. They arrived and repaired the wire, ending the problem.

Aluminum siding can become charged by a damaged or misplaced wire. The right conditions can deliver a deadly shock to a firefighter touching the siding or placing a metal ladder against the house. The hazard would be increased by standing on a wet surface. Unseen arcing of the siding against a metal object could cause some combustible to ignite under the siding, unnoticed by firefighters. This fire could smolder unseen and later break out as a full-fledged structure fire. Aluminum siding isn't the only thing that can become charged. A ground wire, improperly attached to the gas piping, could cause a spark as an unsuspecting plumber worked on the gas line. The spark could ignite gas. In one case I know of, a misplaced wire charged sprinkler piping and all of the metal objects it contacted. The resulting sparks ignited fires in interconnected buildings and resulted in several hectic hours of searching for and extinguishing fire.

WIRES DOWN

I remember once responding to a report of wires down. It was during a torrential downpour, and through the rain-streaked windshield of my chief's car, I could see that the first-due engine and ladder were already on the scene. They had taken folded lengths of hose and laid them on top of several power lines that had fallen into the street. One downed line was arcing and whipping against the side of the pole. As I got out of the car, the truck lieutenant came up to me and reported that one of his men had been hit by a falling wire and was being treated by EMS personnel at the scene. All of the wires except the arcing one were secure, all civilians had been removed from danger, and the utility company was responding.

Sometime after the incident, I spoke to the injured firefighter, and he explained what he remembered of his brush with a high-tension wire. He explained that, when they arrived, a few wires were down in the street, and

Folded bundles of 2½-inch hose have been placed on these downed wires to prevent them from whipping around. The apparatus has been spotted in a safe location, and the danger area has been evacuated, pending the arrival of the utility company.

there was arcing up on the pole. He had been ordered to remove civilians from the stores under the power lines and then to tape off the area with fire line tape. He and his fellow firefighters were just finishing up cordoning off the area with tape when he heard a loud boom. Looking up, he saw a hanging power line whipping against the building. At his periphery, he saw everyone beating a hasty retreat. Before he was able to react, he felt what he described as a wind pass through his body, starting at his extended fingers. He knew he should be running, but his body wouldn't move. He tried to call for help, but he was unable to speak. Finally, his legs gave out and he fell to the ground. When he fell, luckily he fell away from the charged wire and was freed from the deadly current. He was treated on the scene and then moved to the burn center for three days of observation.

He was lucky. The current didn't pass through his heart or any other vital organ. It entered one side of his finger and exited the other, leaving an entrance and an exit wound. The wire was carrying 27K volts. It easily could have killed him. Fortunately, he is back fighting fires, none the worse for wear but having gained a healthy respect for downed wires.

22 • Responding to "Routine" Emergencies

This arcing wire eventually ignited the pole. The area was taped off and personnel were kept at a safe distance until the electricity could be curtailed.

ON-POLE EMERGENCIES

Pole-Mounted Transformers

Pole-mounted transformers pose the same problems as do ground transformers. They can leak coolant oil as a result of damage, faulty bushings, age, or overheating. The overheating oil expands and is forced out of joints in the transformer, usually at the bushings at the top. This oil can contain PCBs. A transformer can explode, spewing flaming oil and fragments to the surrounding area. The explosion can cause live wires to fall onto the ground, nearby buildings, or other wires, charging them. It can also cause the pole to ignite.

A basic rule to follow is not to spray water onto a burning transformer or a burning pole near a transformer or electrical wires. The pole is thick and will take a long time to burn through. The cross beam supporting the wires isn't as thick and will burn through more quickly. Although allowing the cross beam to burn would eventually cause it to fall, taking wires with it, applying water may cause an arc that will instantly drop the charged lines into the street. The water applied to a damaged transformer or arcing wires may cause a short, which in turn can cause an explosion. This explosion may compound your problem. A burning transformer may rupture. A burning or smoldering wire may break and drop to the ground. Then you would have a downed wire to contend with in addition to the burning transformer, as well as the possibility of PCB contamination spread by the explosion. It is better to wait for the utility personnel to cut the power before applying water to the burning pole. In addition, depending on the distance and the voltage present, there may be a danger of electrocution to the firefighters operating the hose. What you think is a low-voltage wire may in fact be in contact with a high-voltage line, making it deadlier than you suspect.

Restoration of Power

Treat All Wires as if They Were Live. Many overhead wires have automatic lockout provisions—if a circuit breaker trips on the pole or in the plant, it should cut the power to the line. The problem is that the system is also set up to restore the power automatically, and it will typically try three times to restore power to the line (three trips to lockout). When you arrive at the scene, you have no way of knowing whether the power has been restored. A seemingly dead line can suddenly become energized. A car driven by a woman struck a pole and, as a result, the overhead power line dropped over her car. She exited the car safely but returned to retrieve her purse. When she touched her car, she was electrocuted. The line had been deenergized when she left the car and then energized again as the power was automatically restored. Another hazard to consider is that both sides of a downed line may be charged, since the power grid may supply a wire from more than one direction. It is also possible that a downed line, even after the power has been removed, can contain dangerous electric potential that must be relieved before the line is safe to touch. Treat all lines as if they were live. When wires are down and you must work near them, put something heavy on them to keep them from whipping. We have used rolled-up lengths of hose successfully at such incidents. This is, however, an inherently dangerous act, since to place the hose, you must get near the wire. If the benefit gained isn't worth the risk, don't do it. If you can remove everyone from the danger area, then there may be no need to weight the wires. On the other hand, if the wire is blocking your path into a burning building with a life hazard inside, then it may be worth the risk.

DOWNED POWER LINES

Vehicular Causes

An automobile accident can snap a pole, and a careless truck driver can snag a low-hanging wire. Either might cause the entire pole to fall or may snap the lower portion of the pole off, leaving the upper portion swinging, supported by the wires. The wires aren't meant to take this stress and could break, dropping everything to the ground. The result would be multiple wires down, and if the pole held a transformer, there would be the danger of an explosion and leaking oil. The wires could also contact each other, causing normally safe wires such as cable TV and telephone lines to become charged. In any case, the fire department will respond and be faced with the possibility of both human injury and fire.

Responding to a pole hit by a vehicle could entail victims trapped in a car that has been energized by the fallen wires. This compounds the problem. There could be injured victims in need of extrication and possibly a vehicle fire, but personnel would be unable to approach the car because of the downed wire. If you were to touch the charged car, you would provide a path to ground. If a victim were to step out of the car, he could create a path to ground and be electrocuted.

Whenever possible, wait for the utility company to remove power from the downed line. If the victim is in need of immediate assistance, you can remove the wire from the car, but personnel must exercise extreme caution in doing this. Since you must avoid direct contact, one method of removing the wire is to throw a length of rope under the wire to a firefighter on the opposite side of the vehicle. This firefighter then tosses the rope back over the wire to the first firefighter, who pulls the wire off of the car by pulling both ends of the rope. Another method is to pull the wire off with a wooden pike pole. Use as long a pole as is available. Neither method is to be considered safe, and you must only use them

Since there is no immediate life hazard within the car, there is no reason to remove the live wire draped across it. The area has been taped off by firefighters, who await the arrival of utility personnel.

when life is in danger. Even a seemingly dry rope or pike pole can contain moisture or contaminants that might conduct electricity. The firefighters should be standing on dry wood and wearing lineman's gloves, if available. Once you remove the line, you must secure it so that it doesn't whip, possibly contacting someone or something. Never move the wire unless life is in danger and there is no unnecessary risk to the rescuer.

Weather-Related Causes

Lightning. A lightning strike could have the same effect on a pole as a vehicle mishap. Wires could come loose, the pole could be knocked down, or the pole could ignite. The resulting fire could burn through the wire, the cross beam, or the pole itself. A lightning strike to a transformer or to the wires could cause a fire, or the wires could break. It could also send a spike of electricity down the service wiring into a building. What could at first seem to be a wires-down incident might actually entail an electrical structural fire in a nearby building. You must search the buildings that might be supplied with electricity from the pole to determine whether their electrical systems have been affected.

One lightning strike to a transformer that I responded to resulted in a two-alarm fire at a nursing home. The surge of electricity sent to the building's service panel in the cellar ignited wires in the panel, and the fire spread to nearby combustibles. The burning electrical wiring produced dense black smoke that permeated the building, requiring the evacuation of the home's seventy-two residents.

Wind and Ice. Wind can cause wires to sway excessively and eventually snap. It can also bring down trees and branches. Not only will charged wire be down, but even parts of the tree can be electrically charged. Trees can conduct electricity, especially if they have been soaked by a recent rainstorm. A freezing rain can build up a thick layer of ice, the weight of which can bring down trees, branches, or the wires themselves.

Salt Spray. In areas near the ocean, salt spray can build up on electrical wires, transformers, and connectors. When it rains, a conductive path is made between the wires, resulting in an electrical arc that can break them.

Repeated Temperature Changes. Constant, prolonged temperature extremes can cause connections to loosen. Loose connections can heat up and ignite the insulation, causing the lines to fall.

Prolonged Hot Spell. This will create a demand for electricity. If the demand exceeds the capacity of the wires, the wires can heat up and fail.

Exposure to Fire. A fire underneath wires can cause them to fail. A rubbish fire, brush fire, car fire, or structural fire can bring down power lines, adding an electrical emergency to the original problem.

A house fire caused these wires to ignite, adding the possibility of downed wires to the hazards. (Official FDNY photo.)

Wildlife. Even nesting birds can cause trouble. A family of birds once took up residence in a pole transformer. They collected twigs and leaves and constructed their nest. All was well until a fire broke out in the transformer, probably as a result of the nest interfering with the cooling process.

This poletop bird's nest could easily cause the transformer to overheat.

Responding to Downed Power Lines

Locate the Hazard. When you arrive at the scene, first locate the break. If the incident occurs at night, illuminate the area with whatever lighting is available. Once you locate the break, you must determine which wires are down and where they are lying. Exercise extreme caution when you leave the apparatus. If a line is on the apparatus and you step off of it, you may complete the circuit to ground. Also take care that you don't step out of the truck either onto a wire or into a puddle containing a wire. If you notice a tingling sensation in your feet or legs, you may be standing near a downed high-tension line. Do not park the apparatus under overhead wires. A seemingly safe location may become hazardous if the incident escalates.

A broken wire has two ends. Once you have located the downed wire, try to determine whether the other end is also on the ground or lying across some elevated obstruction. One end may be too short to reach the ground, but it may be hanging low enough to be a hazard to anyone walking underneath it.

Remove Civilians From Danger. When encountering downed wires, remember that your primary concern as a firefighter is to save lives. You must remove civilians from danger and prevent them from entering the danger area. Often the danger area can be larger than first anticipated.

It may not be necessary for a person to pick up a downed wire to be injured. A downed wire can suddenly snake on the ground, possibly contacting someone standing nearby. Aside from the dangers of puddles and charged vehicles as mentioned above is the chance that a supposedly safe wire is, in fact, charged. In one case, a utility employee picked up a downed power line that was supposed to be uncharged. Several blocks away, however, another downed line came in contact with the same wire, resulting in his electrocution.

Set Up a Danger Zone. The danger zone should be at least one span, or the distance between two poles, in either direction of the fault. This can be quite a long distance. At least initially, the firefighters may have to enter this danger zone to search for civilians in need of first aid or escort to safety. In a populated area, the recommended danger zone can contain hundreds of occupants who may have to be evacuated. Exercise extreme caution when entering the danger area. Consider using the rear entrance of a building as the point of ingress and egress. Doing so may keep you and the evacuees out of the danger area. It may be safer to have the occupants of nearby buildings remain inside rather than let them pass near or under severed wires.

At one incident, an urban sharpshooter testing out his new revolver shot and hit an overhead high-voltage line, causing it to drop into the street. We arrived and taped off the danger area with fire line tape, stood back, and waited for the utility company to arrive. As we waited, a party in a nearby building ended, and several hundred teenagers attempted to exit the building, which was

28 • Responding to "Routine" Emergencies

The danger zone at this corner location could extend in all four directions. How many types of wires can you identify strung along the top of the urban forest?

located in the danger zone. Police officers on the scene had to keep the revelers in the building until the power was removed.

The danger zone may be larger than you think. If the charged power line is lying on a metal fence, the entire run of the fence may become energized and dangerous. This may present us with more than an electrocution hazard. A fire department responded to a report of a fallen power line. When members arrived, they found a line down, roped off the area, and then took up. They didn't notice that the line was on a metal fence, which was energized and heating up. A house to which the fence was attached caught fire and suffered extensive damage. In all cases of downed electrical wires, you should remain until the utility company arrives, if possible.

In another incident, a downed power line caused a short in an electrical outlet in a residence. Sparks from the short fell through the electrical chase into the cellar, where they ignited combustibles in storage there. Two firefighters died fighting that cellar fire. Always consider the possibility that a downed wire or malfunctioning transformer might provoke an interior structure fire. You must check the surrounding buildings for odors, heat, smoke, and fire.

electrical MANHOLES

One day, years before I considered joining the fire department, I was walking down the street toward a gathering crowd. Lights were flashing on several red trucks. Flames were leaping up from a manhole in the middle of the street as firefighters cautiously approached with their hose. As the firefighters poured water onto the flames, several loud explosions shook the ground, and a spectator screamed, pointing to the overhead wires. We all looked up as an overhead wire dropped, flaming and arcing into the street. Other wires were flaming and threatening to drop onto the crowd. I was still half a block away when the panicked crowd turned en masse and started running toward me. I was reminded of a B-grade horror movie in which a monster chases the populace through the streets. At first I stood and watched, fascinated by the scene. When I realized that the crowd was stampeding in my direction, instincts of self-preservation took over, and I, too, began to run.

I would recall this incident many years later when I was studying my department's procedures for manhole fires. The SOPs warned against putting water into a manhole unless so directed by the utility company. I wondered then whether this incident had been the reason the SOP was written, or whether the SOP had been in place and simply ignored.

As a firefighter in Manhattan, I responded to many manhole fires. When requested by the utility, after they had curtailed the power, we would flood the manhole. We wouldn't, however, stand there holding hoses while flowing water into the flaming or smoking orifice. We would be instructed by our officer to lay the open butt of a hoseline near the opening. It would then either be tied off or a heavy object would be placed on top of the butt. All personnel would then be pulled back a respectable distance before we'd start the water.

Some time ago, I learned of an incident in which firefighters attempted to flood a manhole at the request of the utility company. These firefighters stood holding the hose at what they felt was a respectable distance. As the water entered the opening, a deafening explosion ensued. The street was pushed up as if in an earthquake, and all of the firefighters were knocked down. Some were burned; others suffered trauma injuries from the fall. Several never returned to active duty. The pump operator, thinking they had all been killed, transmitted a panicked radio message requesting help.

I have to wonder whether the officer had conducted a risk-to-benefit analysis before ordering his personnel to apply water down the manhole. What was to be gained by holding the hose rather than placing it down and backing away? The only life hazard at this time was that of the firefighters. What property were they protecting? In most cases of fire in an electrical manhole, the utility company must replace the contents of the manhole as well as much of the wire leading into it. Often they will have to replace all of the wire between adjoining manholes. Perhaps the officer just didn't realize the dangers involved in electrical manhole fires.

What Is in a Manhole?

Does a manhole contain electrical wires, a transformer, telephone lines, sewage, or a gas line? There are different types of manholes, the most common being electric utility, gas utility, telephone company, water supply, sewage, and steam supply. At any incident, you need to recognize the type of service that you are dealing with before you decide on a course of action.

Many manhole covers are inscribed with identifying marks, and some have distinctive shapes. The entire name of the associated utility may be inscribed on the cover, or it may only have a coded number. In some cases, its specific shape or design can be used to identify it. You must learn to recognize and identify these covers and associate them with the proper utility to determine their purpose. Once you have identified the appropriate utility and the purpose of the manhole, you can assess the risk associated with it.

Electric Utility Manholes. There are several types of electric utility manholes, based on what equipment is present and the voltage involved. Where overhead wiring isn't evident, the supply to a given building will be underground. In some areas, there is a combination of overhead and underground wiring. In other areas, you won't find underground wiring or electrical manholes, only overhead wiring. You should know which category your district fits into well in advance of your next response.

One type of electrical manhole is used for wiring only. It can contain either distribution lines or service lines or both. In my district, the distribution lines carry voltages as high as 27K. The service lines carry current that has been stepped down to 124 volts by a transformer, which may itself be underground in a manhole, for supply to residential customers. Some manholes carry both service and distribution voltages. The higher-voltage wire is typically located farther away from the street level or, in other words, deeper in the hole, and it is heavily insulated. That doesn't mean that the wires near the top of the manhole are safe. All wires in an electrical manhole should be considered deadly.

Transformers. Another type of electric vault contains a transformer, used to step up or step down voltage. Before residential use, that stepped-down voltage is approximately 124 volts. For heavy users of electricity, the stepped-down voltage will be in the vicinity of 240 volts. These voltages may vary depending on your local utility.

As mentioned above, a transformer consists of a series of wires wrapped around a metal core that is sitting in oil for cooling purposes. The oil in these transformers is flammable and may contain PCBs.

Underground transformers have gratings over them—not just one or two gratings, but five or six. The gratings that I am familiar with cover openings of four by six feet or greater. One or two gratings can indicate a service hole, containing wire going to nearby buildings, but not ordinarily a transformer.

When all is working well, transformers pose no danger either to firefighters or the public. When one is involved in or located near a fire, however, we must maintain prescribed clearances from it with our hose streams, tools, and ladders.

Electrical Emergencies • **31**

The contents of a manhole may or may not be apparent from its cover. Still, you'll need to know what type of service you're dealing with before you can decide on a course of action.

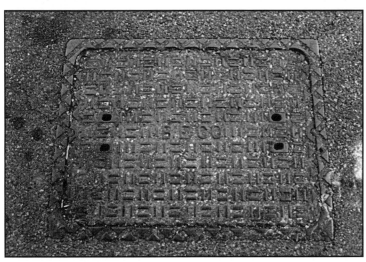

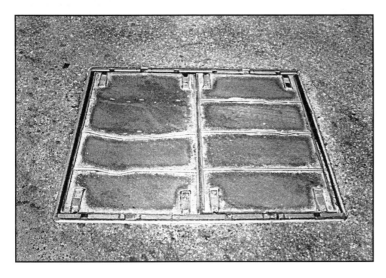

A fire that involves a transformer can be a deadly circumstance. Consider the following:

A woman was pushing a baby stroller over a grating that covered a transformer. As the stroller passed over the grating, the transformer blew up, spewing flames and fragments of metal, killing the baby. In another incident, a woman looking out of her fifth-floor window reported seeing orange flames flaring up *above* her window when a transformer under the sidewalk exploded. There had been a problem with a specific transformer being supplied to the utility by a specific manufacturer. The utility has since removed these transformers from service, but incidents such as these should serve as warnings to fire department personnel that an underground transformer explosion is a dangerous event. Since the covers of these transformers are gratings, much of the explosive force will be vented and, as a result, the gratings won't fly as high as a manhole cover; still, parts of the transformer will be blown out of the hole. The obvious lesson is never to stand on the grating or near a transformer that is smoking or exhibiting some other sign of trouble. Instead, tape off a perimeter at least six feet from the opening, attaching the fire-line tape to cars, fences, garbage cans, or whatever else is available. Newer transformers are designed to minimize the danger of explosion. Their design is such that the transformer will fail at the ends, thus minimizing the flying debris. However, when you arrive at the scene of a fire or other emergency, you have no idea what type of transformer you're dealing with, and there is no guarantee that even the newer type will fail as intended. Stay away from smoking or overheating transformers and wait for the utility emergency crew to arrive.

The emergency crew may ask you to flood the transformer, but you should not undertake this of your own initiative. Some transformers are submergible and some are not. If you flood the nonsubmergible type, you run the risk of allowing PCB-contaminated oil to escape.

Transformers heat because of overload. This is most common during periods of great demand, such as a prolonged hot spell. It can also occur if several feeder lines supplying an area are lost. The remaining lines must then carry the entire load, placing an overload on those transformers serving the remaining lines. As the transformer overheats, the oils within it expand, eventually causing failure.

Often rubbish accumulates around an in-ground transformer. This rubbish can be ignited by any number of means. The fire department will be called to the scene for smoke from a manhole. Although such a rubbish fire can be handled by a water extinguisher, it will require that firefighters place themselves in a hazardous position over the grating. You should wait for the utility company to respond before taking such action. They will be able to enter the manhole and determine whether the smoke is from burning trash or whether there is some problem with the transformer. On their say-so, you can then use a water extinguisher or hose stream to put out the fire. Often the utility worker will extinguish the rubbish fire himself.

The dangers of aboveground transformers range from electrocution to PCB contamination to explosion. Expert assistance of the utility company is required before intervention. (Photo by Matt Daly.)

Problems With Electrical Manholes

Corrosion. The major problem that we experience with electrical manholes in the Northeast is deterioration due to salt. In the winter, salt is poured onto the city streets to melt ice. This makes the roads passable but wreaks havoc on the wiring below. The salt-and-water cocktail seeps into the manholes, coating wires and transformers, deteriorating the hardware. Salty water is a good electrical conductor and can provoke a short circuit. As a result, the insulation can overheat, burn, and decompose, causing power loss involving large areas. The insulation can burn in subsurface ducts to the next manhole, up the overhead pole to overhead lines, or into a building that is supplied by subsurface wires emanating from the vault.

Noxious Fumes. Burning or decomposing insulation gives off copious amounts of noxious black smoke. Because the underground burning is taking place in an atmosphere of reduced oxygen, large quantities of carbon monoxide are present. Consider the following incident.

The fire department was operating at a manhole fire. Down the block, members noticed a flurry of EMS activity. Medical personnel were carrying sick residents from an apartment building. The occupants had been overcome by noxious fumes.

On learning this, the fire department incident commander had the air in the building tested, and high levels of carbon monoxide were found. At this point, the fire department became involved in the incident. Firefighters proceeded to evacuate the building. They forced doors and pulled unconscious residents, clad only in night clothes, out of their beds and into the street.

One of the people removed from the building died of carbon monoxide poisoning, and twenty-five others were hospitalized. The fifty-year-old conduit running from the manhole to the building was made of wood fiber, and it was smoldering. As a result, large amounts of CO had seeped into the building.

The concentrations of carbon monoxide generated by a manhole fire can be quite high. It isn't uncommon for readings of CO in a smoking manhole to exceed 2,000 parts per million. How high above that number the levels may reach is anyone's guess. My department's meters max out at 1,000 ppm, and the utilities' meters don't read past 2,000 ppm. I suspect that much higher levels are reached, and these high levels of carbon monoxide can infiltrate nearby buildings.

The gases generated by burning wiring are not only noxious and deadly, they are also flammable and explosive. Given the proper conditions, these gases, puffing or seeping out of a manhole, can ignite or even explode without warning. A hole that is pushing heavy smoke may never do more than that, while one that appears dormant can unexpectedly explode, sending the cover sixty feet or more into the air. Some manholes are sealed, and the lack of a vent will allow a smoldering fire to go undetected. At any time during an incident, such a manhole can blow its cover.

Flying Manhole Cover. A flying manhole cover is a significant threat. Electric utility manhole covers can weigh more than 300 pounds. These covers have been known to fly into the air and embed themselves on end in the asphalt street. They have embedded themselves in the hood of a passing car and have been found on the roof of a nearby six-story building. It has been reported that they have flown up to one hundred feet into the air.

How far away from a manhole must you stand to be safe from a flying manhole cover? There is no adequate answer to this question. If you back up to the next manhole cover, are you safe? Not if that's the next one that explodes. Prudent action dictates that you move away from the involved hole while taking care to avoid any others in the area. This includes all types of manholes. At one incident, high concentrations of explosive carbon monoxide were found not only in the burning electric utility manhole, but also in the nearby telephone and gas manholes. When a probe was placed into the ground, high concentrations were found there as well. The traffic control boxes and light pole access panels were blown off of the light poles more than a block away from the smoking manhole. Carbon monoxide had built up inside the box and was ignited by an electric spark generated by the operation of traffic control equipment. At such an incident, you must consider the danger area to include anything that has a connection to the manhole. At minimum, you must locate and identify the next similar type of manhole on either side of the involved one, and keep people and cars away from it.

Some will warn you not to run from a manhole cover, but rather to watch it and only run if threatens you. Others will tell you that, if a manhole does explode, sending its cover into the air, you should move quickly to an area of safety. Still others will tell you to watch it if you can while running in the direction opposite its direction of travel. Following any of the above advice is easier said than done. In all likelihood, the sudden, startling sound of the explosion will initially invoke a reflex reaction from you. I have looked back and run from the area. I have ducked into a building at the sound of the explosion. Normally I just startle and jump up an unbelievable distance and look at the manhole cover only as an afterthought. Expressed simply, the prudent thing is to keep your distance and don't bring your apparatus near the danger zone. Don't park on a manhole cover even if it is a block away from the incident. Remember, if one cover blows, others can blow as well. I have seen manhole covers blowing into the air one after the other for several blocks in a row.

Responding to Manhole Incidents

Contact the Utility Company. Contact the related utility and give them an accurate description of the incident. It is important to relay the exact location of the involved manhole to the utility. This will allow them to identify what type of vault it is and to dispatch the required equipment to the scene. If the conditions are such that your presence isn't required or if you're called away to another scene, utility personnel will need precise instructions to locate the site if they arrive after you've departed. Taping off the scene with fire line tape can serve both to restrict public access to an area of possible hazard and to identify the trouble hole to the utility worker when he arrives. If manholes belonging to other utilities are near the trouble hole, notify the appropriate companies. They will probably want to respond to check the situation. A fire in an electrical manhole might be fed by a leak from a nearby gas manhole.

Determine the Extent of the Problem. This requires a visual survey of the obviously involved vaults, as well as a perimeter survey to locate others that might possibly become involved. Cars may be parked on top of manholes, making this task difficult. Take care to observe any overhead wires present. There may be a connection between the manhole and overhead wires, and trouble in one can affect the other. If overhead lines are also involved, the incident at once becomes more complex, since the source of danger is now both above and below ground.

36 • Responding to "Routine" Emergencies

Firefighters stand by as utility workers deenergize a downed power line. (Official FDNY photo.)

Move Cars. If a vehicle is parked on an involved manhole, do not allow anyone to enter it. Cars have been flipped over by exploding manholes. Remember that the smoke and gases escaping from the manhole may be explosive, and a spark generated by an auto might ignite these gases. Anyone in the car would be engulfed in flame as well as being directly in the path of an explosion. Instead of allowing a firefighter or civilian into the car, have it removed by tow truck, conditions permitting.

Evaluate the Degree of Involvement. Is the manhole seeping smoke, or is it being forced out under pressure? Is the cover blown? Are flames coming from the opening? Are the surrounding buildings involved? Remember, you cannot measure the danger by the amount or color of smoke that you see. Conditions can change drastically in a short time, covers may not be vented, and what appears to be an uninvolved manhole can suddenly blow its lid.

Accurately Report the Conditions to the Utility. During times of high incidence, the utility will likely prioritize its response to emergencies. Accurately describing the incident will allow them to respond to the most serious problems first. Is smoke seeping out of the manhole cover, or is it pushing out under pressure? Are flames visible? Has fire extended from the manhole into a nearby building? Are overhead wires involved? Reporting the answers to these questions and noting any unusual or dangerous conditions will supply the utility with the information needed to assign priority. The situation should be updated as conditions change.

Check Adjoining Buildings. Any building that might receive electricity from the manhole must be checked and then monitored until the incident is over. A building's electric service can heat up and ignite nearby combustibles. Gases that are both combustible and toxic can seep in via the ductwork. These gases can ignite or explode. Deadly carbon monoxide can seep in, endangering both occupants and firefighters alike.

A CO meter is a must at all such incidents. I responded to manhole incidents for many years, but until learning of the wood conduit incident described above, I never worried about CO at manhole fires. Since the wood duct incident, I now routinely check for carbon monoxide and have found deadly levels at several incidents. You must periodically check the surrounding buildings for a buildup of deadly CO and evacuate as necessary.

Set Up a Danger Zone. Set up a danger area and keep both civilians and firefighters out of it. It may be necessary initially to enter the danger area so as to remove civilians from it. Consider evacuating them via the back entrance of a building if this will divert you from the danger zone. Mark the danger zone with fire line tape, and keep passersby as well as the curious outside of the area. This isn't always easy. People often neither notice the tape nor grasp its meaning as they duck under it on their way down the block. It may be necessary to use fire department apparatus or to elicit police help to block the street off to vehicular traffic. An airborne manhole cover will be as deadly to the driver of a passing car as it is to a pedestrian.

Monitor the Manhole. When monitoring a trouble manhole, monitor the ones on either side of it as well. Any of these holes may blow due to a buildup of pressure or the ignition of gases formed by burning wires, or of natural gas that may have seeped into the vault. Some manholes won't be pushing smoke and won't give any indication of potential danger. They are sealed and don't allow gases to escape. The pressure can build up unnoticed in these holes, and it can even push up the surrounding ground if the entire metal casing of the vault is ruptured. Telephone manholes in my response area are often tightly sealed, and the large cover can weigh as much as 1,000 pounds.

During a time when numerous manholes are popping, your resources may be taxed. A manhole may initially explode or arc and be pushing smoke, but then it may suddenly go dormant. That doesn't mean that it is safe. Ideally, you should tape it off and monitor it until the utility company can check it out and deem it safe. That, however, may not always be possible. If your resources are stretched thin, a single firefighter may be left to monitor the manhole while the rest of the company remains in service. The firefighter left behind must have a way of contacting the dispatcher and should call for help should the situation become more severe. When the utility company arrives and declares the manhole safe, the firefighter can call to be picked up or he may return to the fire station on his own. Before leaving one or two firefighters behind, consider their safety. Is the neighborhood such that they might become crime victims? Will they need shelter from the weather? Do they have a ready and available means of communication?

If you're unable to leave a firefighter to monitor the situation, you can ask a responsible civilian to call the fire department if he notices any change in conditions. Although this isn't a good solution, it may be the only viable one during a time of multiple emergencies.

Position Apparatus at a Distance. Consider the possible presence of manholes when you position the apparatus. As mentioned above, prudence dictates that you position apparatus remote from the involved area and not on a manhole. Pumpers should take a tentative position at a safe hydrant. If an electrical fire in a manhole extends to a nearby building, you'll have to stretch lines and extinguish the fire. The apparatus pump operator should stand ready to charge a line or reposition his apparatus as need dictates. Consider stretching and flaking out hose in a safe area in anticipation of possible extension.

Position ladder apparatus out of the danger area. Consider the possibility that you might need to ladder an involved building. Take up a position that provides easy access to the area.

Do Not Pull the Cover From a Manhole. At one time, we routinely pulled the covers from smoking manholes so as to relieve the pressure within. Firefighters have been hurt doing this, and it isn't a good practice. This task should be left to appropriately trained utility personnel. A firefighter who is close enough to pull a cover is too close to the manhole. An explosion can result from the influx of oxygen to a smoldering environment or from a spark in an explosive atmosphere. The firefighter is in danger of being knocked down, burned, smashed by the flying cover, or peppered with airborne rocks and debris. Pebbles, propelled by the explosive force, will come at him like bullets. Utility personnel carry a blast mat that they can place over the cover before they attempt to pull it. This mat has holes for the insertion of tools, and it offers some protection to the workers, but even trained personnel are at risk when performing this dangerous task. At one incident I know of, several utility workers were sent to the hospital with facial burns when gases in the manhole burst into flame as the cover was removed.

Flood the Manhole. When requested by the electric utility, flood the vault. Once a transformer is involved in fire, it will likely have to be replaced. Fire in an electrical vault will have damaged the wires to such an extent that they'll all have to be replaced. In fact, the wire will probably have to be removed all the way to the next manhole. By the time fire personnel arrive at the scene, the damage to the transformer or wires will have been done and there will be nothing worth saving in the vault. Adding water to the offending manhole is a dangerous proposition, and there is little to be gained from it.

Even when the utility company requests that you flood the vault, in no case should you allow your personnel to hold and operate the hose while doing so. The result of flooding a manhole cannot be predicted. The added water might quietly quell the flames. On the other hand, it might result in an explosion that can buckle the ground, and the attendant shock wave could knock down everyone standing in the area. Normally, before requesting that we flood the manhole, utility workers will curtail the electricity by cutting wires in the adjoining vaults. It's possible, however, that the trouble hole is fed by yet another unidentified source that hasn't been cut. The actual wiring of the manholes doesn't always agree with the blueprints. Depending on the voltage present in the line, there may be a significant shock hazard to fire personnel. Another possible outcome of adding water to a manhole is a ball of fire that could engulf and burn the unsuspecting firefighters.

The appropriate way to flood a manhole is to use an unmanned, open-butt hoseline as described above. This hoseline can be supplied by a nearby hydrant, if one is available. A $2^1/_2$-inch or larger line works well for this purpose. There is no need to use high pressures that might cause the line to whip around. Low pressure will do the job. You can tie off the line or place a heavy object on it to keep it from moving. Keep the butt as far away from the opening as practical. The farther away you stay, the safer you'll be. Another option is to open a nearby hydrant and allow the water from it to flow into the hole. This may not always be possible, of course, since success depends on the location of the hydrant and the grade of the street.

Terminate the Incident. The incident isn't over until the utility cuts the supply of power to the hole and you have checked all of the adjoining buildings for fire, smoke, and carbon monoxide. This might mean having utility personnel cut wires in several manholes that feed the involved hole. If the electric service in a structure is heating up, the utility may cut the line that supplies the building.

Once the power has been cut, the fire extinguished, and all of the surrounding areas checked and vented as necessary, the utility company may declare the area safe. After investigating nearby exposures to confirm that there is no danger, the fire units can take up.

Establish Liaison With the Senior Utility Official on the Scene. Although we are in charge at these incidents, we don't have all of the expertise required to deal with them. Since we must depend on the utility company personnel, it's important that we maintain a good working relationship with them. The fire department incident commander should easily be able to recognize the senior utility supervisor on the scene. By mutual agreement, Consolidated Edison of

New York mandates that its senior supervisor wear a white construction helmet at major incidents. Making him immediately recognizable to the fire chief averts miscommunication and promotes an accurate, steady flow of information, as well as cooperation between the agencies.

At manhole and other utility emergencies, it's best to seek out the senior utility representative on the scene for his wise counsel. He knows his job and the dangers of the equipment, and he has the knowledge to safely mitigate the situation.

SUMMARY

Electricity is used in countless ways to make our lives easier, safer, and more pleasant. Unfortunately, accidents occur, and appliances age and fail. Sometimes electricity is used incorrectly. When electricity is unleashed against its user, the fire department will be called to remedy the situation, whether for an odor of smoke, a sparking outlet, or downed wires. We must realize, however, that we cannot always, without the help of the utility company, solve the problem. Often the best that the fire department can to is to evacuate the danger area and wait for trained utility company personnel to respond. It is therefore vital to develop a working relationship with the utility and to deliver accurate information whenever you call on them for help.

Many utilities provide training as to what firefighters should and should not do at these incidents. No amount of interagency cooperation, however, relieves a fire chief from his ultimate responsibility for the incident. As on all calls, the chief must use his experience, training, and good judgment in conjunction with expert advice given to him at the scene. The safety of the public and of property is his responsibility, and that responsibility cannot be delegated to a public utility employee. If the chief has developed a working relationship with the utility personnel, he'll feel more confident about their advice and more comfortable in questioning them when he disagrees with their recommendations. He must seek solutions from the experts, but he must also critically evaluate the information he's given and determine any adverse implications for those he is sworn to protect.

Electrical Emergencies • 41

STUDY QUESTIONS

1. When responding to an odor of smoke in a commercial occupancy, a telephone exchange, or a building that houses a transformer, what hazard may be present in the burning insulation?

2. At an odor-of-smoke call, a flickering, dim, or nonfunctioning fluorescent lamp can indicate that the _____ is heating up.

3. It is possible, in ballast devices manufactured before 1979, that the capacitor contains a small amount of _____-contaminated dielectric fluid.

4. True or false: Removing a fluorescent lamp smaller than eight feet will stop the flow of electricity.

5. Underwriters Laboratories didn't require thermal protection for approved recessed lighting until what year?

6. Since the smell of overheating rubber mimics that of an electrical problem, you should considering checking _____ at an electrical-odor call.

7. If you suspect that ignited lint is the source of a smoke condition, you'll have to check both the clothes dryer itself and the entire run of the _____.

8. True or false: A loose wire in a junction box may cause overheating in another box, provoking a fire remote from the cause.

9. A noninvasive way to check behind a wall for hot wires, smoldering beams, or active flaming is to use a _____.

10. Should the fire department open sealed electrical boxes?

11. In a residential building, will opening the main switch necessarily cut off all power to the home?

12. True or false: Some new meters may be removed without interrupting the flow of electricity into the building.

13. Does aluminum siding pose a threat of shock hazard?

14. Name four dangers associated with spraying water on a burning pole-mounted transformer.

15. Treat all wires as if they were _____.

16. Is it entirely safe to use a rope or a wooden pike pole to remove a downed power line from a vehicle?

17. When power lines are down, the danger zone should be at least _____ in either direction of the fault.

18. If, at the request of the utility company, the decision is made to flood an electrical manhole, the proper procedure you should follow is to _____.

19. The major problem experienced with electrical manholes in the Northeast is deterioration due to _____.

20. The gases generated by burning wiring are not only noxious and deadly, they are also _____ and _____.

21. The author recommends that you routinely test for dangerous levels of _____ at every manhole incident.

22. Who should pull the cover off of a smoking manhole?

Chapter Two
Home Heating Emergencies

During the fall, fire departments experience an increase in the number of alarms received for home-heating emergencies. This is because home-heating units that have been inactive for the summer are suddenly being fired up at the onset of cold weather. When these units are properly maintained, the fire department isn't needed. It's usually when they're not maintained that we get called.

A home-heating emergency can escalate to a fire if we aren't called in time or if we don't take the appropriate action. Home-heating emergencies can take many forms, so to mitigate them safely, we need a basic knowledge of heating systems.

According to the NFPA, in 1992, home-heating equipment was responsible for 83,400 fires, 489 civilian deaths, and 2,163 civilian injuries. In addition, the Consumer Product Safety Commission estimates that 210 home heating-related deaths result each year from nonfire-related emergencies due to carbon monoxide poisoning. When a heating unit isn't functioning properly, the carbon monoxide in exhaust gases is increased and can spill into the home, rather than rising up and out of the flue as intended.

When you respond to a heating-related emergency, you must find out what type of heating unit you're dealing with. You need to know what fuel is burned, whether oil, natural gas, or propane. Thick, black, oily-smelling smoke is symptomatic of an oil fire. If the type of fuel isn't immediately obvious, you may be able to find out by asking the homeowner. If the unit is in a private home, chances are good that the resident will have some knowledge of the system. Those chances are reduced if the resident is a renter in a multifamily building. It's also useful to know whether the system is a hot-air, hot-water, or steam-heating system. Each type of system and fuel has its own dangers. To minimize the risks, we must know the dangers, how to avoid them, and how to mitigate them.

You'll notice that I didn't mention repairing the heating system. The role of the fire department should be to remove the hazard and then to leave the home in a safe condition. It isn't our duty to repair the system or to restore to service a heating unit that has shut down or malfunctioned. If, when we arrive on the scene, the system has shut down, it's because of a problem with the equipment. This problem may become worse and eventually cause a fire or explosion if not repaired. If one of our members makes a repair and the system malfunctions again, the department as well as the individual firefighter will be liable for any injury or damage that results. Our role should be confined to eliminating the hazard.

HOME-HEATING FUELS

The most common home-heating fuels are oil and gas. There are varying grades of fuel oil, and several gases are used for home heating. The most common home-heating fuel oil is No. 2 oil, while No. 6 oil is often used in commercial applications. Natural gas and propane are the most common home-heating gases.

Fuel Oil

Fuel oil is one of the by-products of the distillation of crude oil. In the distillation process, oil is separated by viscosity into No. 1 (kerosene and jet fuel) through No. 6 fuel oil. The lower the number, the lighter the oil. Light oils have low viscosity and flow easily, whereas heavier oils have high viscosity and must be preheated to flow properly. No. 2 oil is used in home oil burners. No. 6 oil is used in larger burners, such as the ones found in apartment and commercial buildings. Fuel oil has a minimum flash point of 100°F. In reality, the flash point may be as high as 130°F. This high flash point is an important safety factor for oil heating units. Unlike natural gas and propane, leaking fuel oil doesn't pose an immediate fire hazard, since it isn't as easy to ignite.

Fuel oil doesn't ignite in the liquid state; rather, it must first be vaporized. As a result, spilled fuel oil isn't likely to be ignited by a random spark or cigarette butt. If, however, the oil is heated to its ignition point or is contaminated by a low-flash-point product such as gasoline, it can give off ignitable vapors. If the fuel oil is contaminated by heavier oils, its flash point will increase, and it will become more resistant to ignition.

A gallon of oil requires as much as 2,000 cubic feet of air to burn completely. Older conventional-head burners produce a flame that is between 1,200°F and 2,800°F. When fuel oil burns with the proper air-oil mixture and at the proper temperature, it burns clean, creating little smoke. If the flame burns cooler than it should, excess smoke will be generated. Excess smoke is a result of incomplete combustion, and it is accompanied by the formation of carbon monoxide and other toxic gases. Combustion gases that are cooled excessively, as they might be in cold weather, will condense in the flue, and the resulting acidic condensate can cause the metal flue to deteriorate over time. This results in perforation of the metal flue and allows toxic gases to spill into the living space. If cooled enough, the draft can fail and the toxic gases will be spilled into the home rather than exiting the flue as intended.

A home heated by oil will have a fill pipe somewhere on the outside to receive fuel, and there will be a vent pipe, usually somewhere near and extending above

The oil tank vent pipe is collocated with the fill pipe.

the fill pipe. It will also have a fuel storage tank. Typically, in a private dwelling, this will be a 275-gallon tank. In other occupancies, you may have more than one 275-gallon tank, a 550-gallon tank, or a 1,080-gallon tank. A 275-gallon tank is typically located in the building, whereas the larger ones are normally outside or buried in the ground. Some localities require tanks to be vaulted. The type of tank that you encounter and its location depend on the type of occupancy and the local codes.

The purpose of the vent pipe is to let air exit the tank as fuel enters it. Some vent pipes are equipped with a whistle alarm that sounds as the tank is being filled with fuel oil. The escaping air activates the alarm. If the alarm doesn't sound, the cause could be a blocked vent. Pressure buildup as a result of the blocked vent could quickly result in tank rupture and an oil spill.

A home-heating unit usually has a copper fuel supply pipe of $3/8$-inch to $1/2$-inch diameter, running from the tank to the burner. In some installations, steel or iron piping is used. There will be a fuel shutoff valve located at the oil burner and possibly another at the tank. The valve is usually either a wheeled valve or a levered handle. Both should have a fusible link to shut off the flow if the valve is exposed to fire. When a return line is used, the supply pipe is the one that is connected to the filter, a small canister usually located before the oil burner motor. In the case of a furnace converted from fuel oil to gas, the fill pipe may still be present, as may the oil tank. Looking at the furnace itself will conclusively determine which type of fuel is used.

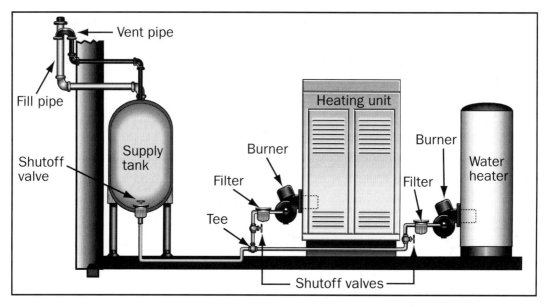

A typical fuel oil system, tank in the cellar. Note the location of the shutoffs.

HOW OIL BURNERS WORK

To ignite the fuel oil in the combustion chamber of an oil burner, the fuel is atomized, reduced to tiny particles, thus increasing its surface area. The fuel-air mixture is then ignited by an electric spark in most home burners, and in some commercial units by a gas pilot or a combination of gas and spark. The domestic unit will consume anywhere from one to ten gallons per hour, whereas industrial types may burn up to fifty. At ten gallons per hour, about five ounces of fuel are burned each minute. Safety devices allow the oil burner fuel pump to operate for no more than ninety seconds if no flame is detected. If there is a problem, the motor to the pump is stopped and fuel oil is no longer delivered into the combustion chamber. At five ounces per minute, this means that about eight ounces of fuel can be pumped into the combustion chamber even if the ignition sequence fails. Even more than that will be delivered in large commercial units. If ignition occurs at a later time, that excess eight ounces will create a smoky fire. If some of that eight ounces has been vaporized by the hot walls of the combustion chamber, an explosion may ensue.

In domestic installations, the oil burner is in operation about eight hours out of every twenty-four. A typical burner operates approximately 1,500 hours a year. With all that use, it's amazing that we aren't called to oil burner emergencies more frequently.

Components of an Oil Heating System

Storage Tank. Local codes dictate how far the tank must be from the heating unit and whether or not it must be enclosed or in its own room, located outside, or buried. Tank corrosion caused by accumulated water can cause failure and a resultant fuel spill. A fuel leak from a buried tank can go unnoticed for years.

Fuel Pump. The fuel pump delivers fuel to the burner at a steady pressure. Home-heating units usually deliver fuel at 100 psi. Newer burners are being designed to operate at higher pressures.

Motor. The motor drives the fuel unit and rotates the fan. Older burners operate at 1,725 rpm. Newer ones operate at 3,450 rpm.

Nozzle Assembly. The nozzle assembly separates the fuel into tiny droplets, approximately 55 billion per gallon of fuel at 100 psi. This increases the exposed surface of a gallon of oil to 690,000 square inches. These tiny droplets are easily atomized as the fuel is delivered to the combustion chamber at a fixed rate and in a cone-shaped pattern. For the fuel to ignite properly, the spray from the nozzle must impinge on the ignition spark at the proper angle.

Fan. The fan delivers a precise amount of air to the combustion chamber, where it is mixed in proper proportions with the atomized fuel.

Combustion Chamber. The combustion chamber is where the actual heating takes place. It is made of steel or fire brick. It's designed to withstand the tremendous heat generated by the oil flame, but it isn't designed to withstand direct flame contact.

Heat Exchanger. The heat exchanger is a series of tubes above the combustion chamber. Hot combustion gases pass through these tubes on their way out of the boiler or furnace and into the flue. These tubes are surrounded by either air (in a hot-air system) or water (in a hot-water or steam system). The heat is transferred from the combustion gases to the air or hot water.

Flue Pipe. The flue pipe is usually a galvanized metal piping system that delivers combustion gas from the furnace to the chimney.

Barometric Draft Damper. The draft damper is a flapper-type door on the draft pipe that swings open and closed with variations of atmospheric pressure. Failure of this door to swing freely can prevent a proper draft from delivering flue gases up the chimney.

Electrical System. Oil burners use a step-up transformer and a combination of electromagnetic valves (solenoids), relays, switches, and resistors to control the burner. The transformer receives energy at 120 volts and produces energy of as high as 10,000 to 12,000 volts for the spark that ignites the oil.

Remote Control Switches. The emergency burner shutoff switch and the thermostat shut down the system from areas remote from the heating unit.

Thermostat. A heat-activated switch that reads the ambient temperature and then calls for heat or shuts down the unit, depending on the settings. In hot-air and steam systems, the thermostat directly controls the starting and stopping of the ignition cycle. In a hot-water system, the thermostat only controls the circulator motor. When heat is called for, the circulator motor circulates hot water. As the water circulates, it cools. It is the aquastat that starts the ignition system in a hot-water unit. When the water temperature drops to a preset point, the aquastat triggers the ignition sequence in the burner.

Burner Emergency Shutoff Switch. A remote switch used to shut off the burner. It is usually located at the door to the burner room or at the top of the basement stairs. Often the switch has a red switch plate. Another switch may be located on or near the furnace. The emergency shutoff switch may not always be present or, if present, may not be working.

Ignition System. Once the oil is delivered into the combustion chamber and is atomized by the nozzle, it must be heated above its ignition temperature. The heat for this is delivered by a spark from the electrodes of the ignition system. A step-up transformer is used to create the spark. There are two types of ignition systems currently in use. In interrupted or intermittent ignition systems, a spark is present only at the start of the burner cycle. Once the oil has been ignited, the ignition system shuts itself off. In constant ignition systems, the spark ignites the oil and then remains on for the duration of the burner cycle.

Limit Controls. This entails a temperature or pressure-activated switch used as a safety control. The limit controls prevent the water and air from being overheated, as well as preventing overpressurization. The high-limit safety turns off the burner if the water temperature, steam pressure, or air temperature inside the furnace or boiler becomes too high. Residential boilers operate at a maximum pressure of 15 psi for hot water and 2 psi for steam. The high-limit safety protects a hot-air system from excessive temperatures in the furnace or in the ducts. In a hot-water system, it prevents the water from overheating and turning to steam. In a steam system, it prevents the pressure from exceeding the design pressure. The low-limit safety controls the burner, depending on the water or air temperature. It is used to keep the water at the appropriate temperature, and it operates at a lower temperature than the high-limit safety. Sometimes it's used to maintain a minimum level of steam pressure.

Low-Water Cutoff. This prevents the burner from operating if the water level drops too low. It is found in steam systems and in some hot-water boilers. This cutoff can either be mechanical or electrical. If mechanical, a device such as a float is used. If electrical, an electric eye detects the water level and triggers the cutoff. If a hot water system has no low-water cutoff, it can dry-fire if the feed water is cut off. This will cause the boiler to become extremely hot and crack. When the thermostat calls for heat, the circulator will kick in and pump a few gallons of residual water into the boiler. The water will convert to steam, and the resultant pressure may blow the boiler apart.

Pressuretrol. This device measures high and low pressures and will turn the system on or off accordingly.

Aquastat. The aquastat measures the temperature of the water and turns the system on or off at the temperature limits.

Primary Control. The primary control protects the unit from failure of the ignition or flame. It will shut down the fuel pump if it doesn't detect a flame in the predetermined time frame. The primary control supervises the burner. It responds to a call for heat by starting the burner, and it responds to the limit switch by shutting down the burner. It controls ignition and checks for the presence of flame. If it detects a malfunction, it shuts down the burner. The older type of primary control is known as a stack switch, and the newer type is called a visual flame detector.

The stack switch, a primary control located in the flue, senses heat entering the pipe. If it senses heat, it shuts down the ignition sequence and starts the running sequence. If it doesn't sense heat, it will cycle on the ignition sequence. If no heat, a result of no flame, is detected, then the stack switch will allow current to flow for between ninety seconds and two minutes, then it will stop the ignition cycle by cutting off the current. Although this prevents a large buildup of oil in the combustion chamber, it does allow some oil to enter the chamber. For a typical home-heating unit, that could be as much as eight ounces. There is a reset button on the stack switch. If it doesn't sense any flame, then the stack switch shuts down the burner. To restart, press the reset button.

The visual flame detector is a newer type of primary, consisting of a solid-state electronic light-sensing cadmium sulfide cell used to sense the fire. It will shut down the flow after as few as fifteen seconds if there is no flame. This modern primary will allow less oil into the combustion chamber if there is an ignition failure than will the older stack switch.

COMMON OIL BURNER EMERGENCIES

Delayed Ignition

Of all the problems encountered with burners, delayed ignition, also known as puffback, poses the greatest danger. It occurs when unburned atomized fuel oil is ignited at the start of a burner cycle, resulting in an explosion. This explosion can be a small smoky puff, or it can be a blast that blows flue pipes out of the chimney, blows open the fire box door, and sends a tongue of flame out of the furnace and across the room.

The protective relays are supposed to shut down the oil burner if they sense a problem, thus preventing puffback. The stack switch or visual flame detector should shut it down if it doesn't sense a flame. The problem is, these devices don't always work.

The causes of delayed ignition are as follows:

Faulty Ignition. This is a leading cause of delayed ignition. Puffback can also be caused by a faulty pressure regulator that allows oil to drip into the combustion chamber after the burner shuts down. The leaking oil gets absorbed by the lining of the combustion chamber, and it can also pool in the bottom of the chamber. The result of this condition is a smoky burn and the sudden ignition of the pooled oil. This sudden ignition or puffback is accompanied by a low bang or thud. The resident calls the fire department, reporting an explosion in the basement accompanied by heavy black smoke.

Leaking Oil Pump Seal. A leaking oil pump seal can cause excess oil to be thrown into the combustion chamber. This excess oil isn't properly atomized, and the larger droplets can fall to the bottom of the chamber rather than ignite. When the burner cycles on and a hot chamber has vaporized the pooled oil, a puffback explosion can occur.

Improper Atomization of Fuel. In older, rotary cup burners, many of which are still in use, atomization is accomplished by centrifugal force. Most newer burners inject the oil into the combustion chamber in the shape of a fine, cone-shaped spray. In either case, once in the combustion chamber, the atomized oil is mixed with a precise amount of air and ignited. If the nozzle or the rotary cup is damaged or dirty, the intended shape of the oil spray can be changed, possibly causing it to miss the electric arc. When this occurs, ignition will fail. The result will be either pooled, unignited oil or unignited atomized oil in the combustion chamber. Both can cause problems.

The result of improperly atomized oil is that larger droplets enter the combustion chamber. They won't ignite easily and can cause buildup in the chamber. The fuel won't burn cleanly, large amounts of black smoke will be produced, and carbon will build up in the chamber and on the various sensors, eventually causing the sensors to fail.

Other causes of improper atomization are water in the fuel or the wrong fuel grade, a lack of draft, improperly gapped electrodes, or a dirty stack switch. Again, the result will be improper atomization, pooling, smoky burn, and an explosion on start-up.

Hard Starting

Delayed ignition may not happen all of a sudden. It is a condition that can develop over time, because of a failure to correct a hard-starting problem.

Usually an oil burner will first suffer a series of small puffs as it becomes harder and harder to start. As the contacts get dirtier, as the nozzle clog worsens, or as the afterdrip becomes more severe, each start takes longer and becomes noisier. Eventually, when conditions are right, an explosion occurs.

A leaking oil pump seal can throw oil on the floor in front of the burner, as well as into the fire box. Oil at room temperature can't be ignited unless it is first atomized. However, this spilled oil can be ignited if it is near enough to be warmed by heat from the burner. Once it is, a spark from the burner's ignition system or from the motor can ignite the vapors and, in turn, nearby combustibles. You now have a basement fire, not an oil burner emergency. A call for smoke in the basement may turn out to be an oil burner emergency or an oil burner fire. If the fire is contained in the combustion chamber, it is an oil burner emergency. The chamber is designed to contain the oil fire, so a fire in it is usually no threat to the residents or the building. If the fire is out of the chamber, it can be considered an oil burner fire.

Afterfire

Afterfire is the burning of pooled oil in the combustion chamber after the burner shuts down. Significantly, this burning takes place when the fan and ignition mechanism are inactive. The chamber is still hot, so the pooled oil is atomized and continues to burn without the required amount of air. As a result, excess smoke is generated. The result will be thick black smoke exiting the chimney; otherwise, if the burner door is open or the flue has been dislodged, smoke will fill the basement.

Anytime a smoky combustion occurs in the burner, soot builds up, coating electrical contacts and ducts. The soot builds up to the point of burner malfunction, eventually resulting in delayed ignition or other problems.

When confronted by an afterfire incident, remember that the oil that's burning in the combustion chamber isn't a problem. It'll burn smoky, but if the flue pipe is connected, the burner room will be unaffected. If burning oil has leaked onto the floor or into the pit, you can extinguish it with foam or a dry-chemical extinguisher.

Shut off the burner, shut off the fuel, ventilate if necessary, and extinguish any fire. If the fire in the chamber is of such magnitude that it's threatening to extend, then you might need to extinguish it. You can use a foam can or two, or a dry-chemical extinguisher, but it's still possible that the vaporized fuel will be reignited by hot metal. You may have to cool the chamber with a fog stream, but be aware of the dangers of a steam explosion and the resultant spalling fire brick, and don't flood the chamber, since this might cause burning oil to leak out. At once such incident, we were called by neighbors for heavy smoke coming from the rooftop of a private home. Inside, we found an oil burner repair technician and a fire in the combustion chamber of the burner. Flames could be seen at the draft damper. The technician said that he was letting excess oil burn off and that there was no danger. We left the fire burning and told the technician to call us if there was any problem. Later, he called us back. The flames hadn't died down, and he requested that we extinguish them. We did so with a few bursts of spray into the draft damper. We shut the oil feed and the emergency shutoff, and we also turned down the thermostat so that the burner wouldn't cycle on. We checked for extension along the path of the duct but found none, and so we left the technician to his task. This time, he didn't call us back.

Consider the possibility of extension when confronted with an afterfire incident. Ask yourself whether a heated flue pipe might have ignited a beam or a stud. Is fire burning in the chimney, and has it taken hold of the structure? Have any nearby combustibles been ignited?

Pulsation

This condition occurs when the flame jumps away from and then returns to the burner nose cone. It can last a few seconds, or it can occur whenever the burner operates. Another cause can be air in the fuel line. When pulsation occurs, the burner shakes. This shaking can cause the whole house to shake, and it has been likened to a train passing by. It is caused by a combination of poor draft, soot, scale buildup on the flues of the furnace, and/or air in the fuel line. It isn't uncommon in modern, flame-retention burners. In one case, I responded to a report of "shaking walls." The caller explained that the whole house was shaking, and that she thought there was an earthquake. We thought that she was deluded until her furnace cycled on. Then we, too, felt the earth move. We quickly traced the problem to the burner and shut it down. Pulsation can cause the door of the furnace to open. To prevent this, the homeowner might brace the door shut with a two-by-four. Later, the two-by-four might fail, letting the door fly off the burner. Caution is necessary whenever you approach any malfunctioning burner. Consider any bracing of a furnace door to be a potentially deadly situation.

When confronted with a burner pulsation incident, be cautious. If severe, this condition could result in an explosion and explosive failure of the door or burner. Locate and shut off the emergency switch, then shut off the fuel supply. Open the fire door from a safe location on the hinge side. Using a six-foot pike pole will put you out of harm's way should flame blow out the opening or the door be blown off the furnace.

The White Ghost

When I first encountered the white ghost, I had never heard of it. My engine company had responded to an oil burner emergency in an apartment house that had escalated to a full-blown cellar fire. We arrived as the third-due engine. The first two engines had each stretched a 2½-inch line and were trying to get down the cellar stairs located in front of the building. The heat was so intense that they couldn't advance, and they were operating their hoselines into the basement, attempting to cool it down. Hot, black smoke was pouring out of the cellar. As we stood by, awaiting orders, the smoke changed from black to a pearly white. I took that as a good sign, thinking it was steam. The absence of black smoke indicated to me that the fire had been extinguished and that we would all be going home soon. I was wrong.

There were firefighters on the first floor, searching for victims and extension. Another member had just entered a second-floor window from a portable ladder, when there came an explosion followed by orange flames flaring out of all of the cellar windows. All of us in front of the building were knocked down by the blast, and we had to crawl away from the intense heat of the flames.

The firefighter on the second floor was okay. He'd gotten off the ladder and into the window in time. Those on the first floor weren't as lucky. They came tumbling out of the doorway, driven out by the heat. The flames had come up the interior stairs, and most of the inside team had suffered burns. The flames, although reduced in intensity, continued to vent out of the cellar windows as we regrouped our forces and renewed our attack on the fire, which was eventually extinguished.

I had no idea of what had gone wrong at this incident, but found out later that I had been introduced to the white ghost.

The temperature of an oil burner's combustion chamber can be as high as 2,600°F. If ignition fails, the oil may be ignited by the high heat of the chamber. If not, the atomized oil will fill the chamber, flue pipe, and chimney. If the flue piping has been blown down as a result of puffback, or if the combustion chamber door has been blown open, this explosive vapor will fill the burner room or the entire basement. If it finds a source of ignition, an explosion may result.

A myriad of faults can cause an oil furnace fire; however, only certain conditions will lead to the white ghost, which is a rarity. If atomized fuel is being overfed into a hot combustion chamber without being ignited properly, there may still be a positive draft up the chimney, driven by air entering through cracks in the chimney and the firebox. Since the fuel is starved for oxygen, vast amounts of carbon monoxide, oil residue, and dense smoke will opportunistically cling to surfaces along the way, and copious amounts of black smoke will be emanating from the chimney. When firefighters ventilate the basement and spray a fog stream on the furnace, however, the direction of the draft may change. Fresh air will now be coming down the chimney, combining with the combustion by-products that have collected within the system, as well as in the basement itself. At the right combination, and in the presence of a source of ignition, a violent explosion may ensue. The fog stream has actually been a contributor in this scenario, since the atomized water droplets readily entrain vast amounts of fresh air; moreover, the vaporization of the droplets can serve the combustion process. The sudden change in color of the smoke from black to white, as well as the strong smell of fuel oil, is indicative of the reverse draft and vaporization that lead to the explosion.

Vaporized oil is white like steam, but it smells and tastes like oil and can irritate the eyes. It's rare that firefighters encounter the white ghost, but if you do, it'll be an encounter that you'll always remember, and you'll never look at an oil burner in the same way again. At the time that I was initiated by the white ghost, my research turned up only one explanation for the phenomenon, and even then, I didn't recognize that the color change in the smoke had been a danger sign. I hadn't noticed the oily smell and wouldn't have known its significance, anyway. Even today, there are still firefighters out there who've never heard of the white ghost. I periodically bring up the topic at drills and always find someone who's hearing about it for the first time. We must never assume that all firefighters know what we know.

Since the white ghost is a potentially explosive situation, you should treat it as a gas leak. Shut the remote control and, if possible, the fuel supply. Vent the burner room as soon as possible, but don't enter the burner room unless absolutely necessary. Expose as few men as possible. If you do decide to enter the area, do so with a fog nozzle set on a wide fog pattern. This is to protect the advancing firefighters as well as to ventilate the area. Have a backup line available to protect the members who have entered the burner room. Be ready to deal with an explosion, possible partial collapse, and rapid fire extension. If you're in the burner room when the white ghost appears, open up your hoseline for protection and back out immediately.

General Precautions for Oil Burner Emergencies

Always Approach a Malfunctioning Oil Burner With Extreme Caution. The conditions that caused a puffback might initiate another if the burner hasn't been shut down. Stay low and away from the front of the burner. In some instances, if puffback has been common, the combustion chamber door may have been jammed or braced shut by the resident. This could cause catastrophic failure of the burner. Be wary of any such setup.

Always Stay Away From the Front of Furnace Doors. Oil burner repairmen and firefighters have been hurt by delayed ignitions. If the furnace door is open when the burner ignites, a tongue of flame can project from the furnace. If the door is closed, it can be blown across the room by the force of the ignition. Always crouch low when you're in front of the furnace. That way, if any fire shoots out of the opening, it'll be above you.

Don't Close the Furnace Door if You Find It Open. Closing the furnace door can cause severe pulsation. This pulsation can blow open fire doors and cause the vent pipes to become dislodged. In all cases, if you shut down a burner, you must tell the resident what he must do to remedy the situation and not to turn the burner on again until it has been serviced properly. Some municipalities will issue an order to repair the burner, and they'll reinspect at a later date to verify that the condition has been corrected. If the malfunction occurs on a cold winter night, consider relocating the residents to a warm shelter.

Don't Put Water Into the Combustion Chamber. Normally you shouldn't apply water into the combustion chamber. Instead, allow any fire there to burn out after shutting down the power and the fuel supply valve. If you apply water to the chamber and extinguish the flame, you might create a new problem. The oil can be vaporized by the hot combustion chamber walls, and instead of harmlessly burning, it might spread as an ignitable vapor into the burner room. If it encounters an ignition source, there'll be an explosion. Foam can be used for extinguishment and to prevent the generation of vapors, but the extreme heat of the combustion chamber will break it down. You'll have to reapply foam as needed until the chamber cools. It may be preferable at first to extinguish the fire, then cool the chamber with water spray, then cover the pooled oil with foam. The application of too much water to the combustion chamber, however, may result in oil leaking out of the burner.

Any water that's applied to the combustion chamber will be converted to steam, possibly explosively. If water is absorbed by the walls of the combustion chamber, it may turn to steam explosively when the burner is reignited by repair personnel.

In most cases, a foam can or two, or a 20-pound dry-chemical extinguisher, will provide all the extinguishment needed, but since a puffback is a situation that might result in a fire, you should stretch a protective line. A fog nozzle would be useful to vent the area as firefighters move in. Fog, if foam or dry chemical isn't available, is also useful in extinguishing a spilled oil fire. If the oil vaporizes and fills the room with white, oily-tasting smoke, the fog line can be used to ventilate the now-explosive atmosphere. Anyone in the building at the time of this changeover from black to white smoke is in grave danger and must exit the danger area immediately. Open the line and operate it as firefighters back out of the burner room. Stretch a backup line with a straight stream to cover their retreat and to extinguish any fire resulting from an explosion.

Stay Out of the Boiler Room if You Aren't Needed There. Keep the number of personnel involved to a minimum. Keep reserve firefighters at the ready in a safe area to stretch additional lines as needed. Also, don't block the entrance door. It isn't uncommon for firefighters to crowd into the oil burner room or basement to see and be part the action. Crowding around the door unnecessarily puts these members at risk and blocks the path of egress for those who are directly involved.

Remember That Shutting Off the Remote Switch Doesn't Shut Off the Oil. Always shut off the oil valve at the tank and/or burner. Otherwise, flatten the supply line with a metal tool, but don't rupture it.

56 • Responding to "Routine" Emergencies

Tactics at a Puffback Incident

Pay Attention to the Smoke. When you arrive at the scene of a puffback, you'll often see thick black smoke pouring from the location of the oil burner. This smoke, unlike the smoke from an oil burner fire, won't be very hot. Typically, oil burner smoke resulting from a puffback is fairly cool and not too difficult to breathe. Don't be fooled by its relatively nonirritating effect, however. It contains carcinogens from which you must protect yourself. Wear SCBA. In addition to the cancer risk, the emergency may turn deadly to the unprotected firefighter if atomized oil, entrained in the smoke, suddenly ignites. If the smoke is extremely hot, this isn't an oil burner emergency—it's a fire, and more is burning than a little oil pooled inside the combustion chamber.

If the oil burner isn't in a dedicated room, but rather, is open to the rest of the structure, the smoke may be throughout the building. Take note of the chimney. If no smoke is coming out of it, but if smoke is pushing out of the windows, then it's a good indication that the flue pipe has been dislodged.

Shut Off the Oil Burner. The initial fire department action should be to locate the burner switch and shut it off. This is an important safety precaution. It will prevent the burner from cycling on, as well as prevent fuel from being pumped into the chamber.

The fuel shutoff on this oil burner is located just before the filter. Remember to shut off the fuel at the tank, too, if possible.

Stop the Flow of Fuel to the Burner. Once the oil burner has been shut down, locate the oil shutoff valve. It can be located at the burner or at the tank or in both places. If there are two valves, shut both of them. As mentioned above, if you can't operate the valve, you may be able to crimp the fuel line. Take care not to rupture it. Once the fuel has been cut off, any pooled oil burning in the combustion chamber can usually be allowed to burn off harmlessly.

The burner might be in a room separate from the fuel tank. In private dwellings, the storage tank often shares an uncompartmented basement with the burner. In homes without a basement, it can be found in a separate room or in an attached shed. The tank can be inside, outside, or buried in the ground. Look for the second oil shutoff near the burner. If you find it, close it, too.

Ventilate. Ventilate the room and immediately launch a search for victims and extension. If the fire hasn't gotten out of the chamber, it isn't likely that it will have extended, but a diligent search is still required. Check openings in the ceiling and any nearby combustibles. Burning oil may have been splattered, and a nearby combustible might be smoldering. In the heavy smoke of the oil burner, smoldering rubbish could easily be missed. Perform ventilation early in this operation. Because of the lack of heat entrained in the smoke, backdraft isn't a concern at an oil burner emergency. Normally, you needn't break windows to ventilate the area. Opening doors and windows should be enough, but if nondestructive means won't adequately vent the area, then you may have to resort to other methods. Remember, venting the area will remove the smoke and allow firefighters to operate in a smoke-free environment. This means increased operational safety for firefighters. Their safety is always more important than the cost of damage incurred by venting. The proliferation of smoke throughout the building may necessitate ventilating the entire structure, as well as searching it for possible victims.

OIL BURNER FIRES

High heat emanating from the basement or the vicinity of the oil burner is a sign of an oil burner fire. Unlike the smoke of an oil burner emergency, this smoke will be hot, as at a structure fire. This is a structure fire, not an emergency. You'll need more than a few foam cans to control it. Stretch the same line that you would for any structure fire. If spilled oil is involved, a foam line might be required.

Tactics at an Oil Burner Fire

Shut the Remote. The switch typically found at the top of the cellar stairs will cut electrical power to the heating unit, shutting off the ignition and the fuel supply.

Stretch a Line. Oil burner technicians carry dry-chemical extinguishers to use on the oil fires they encounter during the course of their repair work. Dry chemical works well on an oil fire. In most cases, however, the technicians are only dealing with a small, contained oil fire. They don't enter hot, smoke-filled basements as firefighters typically do. They don't need a handline. Firefighters, however, don't have the luxury of knowing whether it is an oil-only fire. Often, by the time you're called to the scene, a small oil fire will have escalated to become a structure fire. Since you won't know for sure which it is, stretch a line appropriate for a structure fire. In a private dwelling, a 1¾-inch line will do the trick. If on entry the fire can be controlled by water, dry-chemical, or foam extinguishers, then the handline need not be operated, but at least it is ready to go into operation should it be required.

Consider Using a Fog Nozzle. A fog nozzle is effective against burning oil. A straight stream, although not as effective at extinguishing burning oil, won't generate the excess steam that a fog stream might. It will, however, be effective at extinguishing the involved combustible contents of the burner room. A combination nozzle offers the best of both types at such a fire by providing ventilation as well as extinguishment for both oil and solid combustibles. Improper use of the combination nozzle can, however, result in steam burns, and it can pull heat and fire onto the nozzle team if they pass a pocket of fire or a hot wall.

Stretch a Backup Line. If the oil atomizes, becomes entrained in the smoke, and ignites, the firefighters will need a backup line to cover their escape.

Vent Ahead of the Line as It Advances. A fog line will push the dense black smoke ahead of the engine company and ultimately out of vented windows. If no vent is available, do not use the fog stream, since steam burns may result. Often cellar areas are difficult to vent. If high heat prevents entry into a cellar, consider cutting a hole in the first floor, near a vented window that's opposite the direction of the advancing line. Protect this opening with a charged line and, if possible, cut the roof above it. This will give the steam and heated gases a way to escape the cellar, making advance easier.

Monitor the Color of the Smoke. Be wary of a change in the color of the smoke. A change from black to white might indicate that the fire is under control and producing steam, or it might signal the onset of the white ghost. An oily odor accompanying a pearly white smoke is indicative of the white ghost. Treat the white ghost as a gas leak.

Shut the Fuel Supply. Always shut the fuel supply in addition to shutting the remote switch. If the remote doesn't shut off the burner or if there is no remote, or if oil is leaking into the combustion chamber, shutting the fuel supply valve at the burner (and at the tank, if one is present) will cut the fuel supply to the burner. If the valve is defective, it may be possible to crimp the supply tubing with pliers or with blows from a metal tool. This will work well if the fuel line is made of a soft metal, such as copper. Take care not to rupture the fuel line.

Make a Thorough Search for Fire Extension. If fire or flames have extended from the fire box, examine all possible avenues of extension. Check the areas above and around the furnace. Remember, burning oil might have splattered out of the combustion chamber, and hot pipes and ducts can char and eventually ignite nearby wooden beams and studs.

Don't Extinguish Fire in the Chamber Unless Necessary. As mentioned above, normally it's best to allow fire in the combustion chamber to burn itself out. However, if the fire in the chamber is threatening to extend to the structure or to nearby combustibles, then it may be necessary to extinguish it. To do so, you must cool the chamber so that it no longer atomizes the remaining fuel, possibly creating a white ghost condition. Putting water into the chamber may result in a steam explosion and possibly splash burning oil. It may also damage the burner, perhaps explosively. If water is absolutely required, apply it sparingly and from a safe position. If you apply foam to the chamber, it will prevent the generation of flammable gases for a time, but the high heat will eventually break down the foam blanket, and the vapors will again escape. If they find an ignition source, they may ignite. You'll have to replace and monitor the foam blanket. If you're using plain water, you'll have to cool the chamber to prevent it from generating flammable vapors. Excess water or foam in the combustion chamber can leak out, spilling flaming oil that will float on top of the water. Dry chemical may be the best choice for extinguishing a fire in the combustion chamber. It will neither create a steam explosion nor splash burning oil, nor raise the level of the burning oil, allowing it to spill out. Some departments carry a squeeze bottle of dry-chemical extinguisher for this purpose. A few quick squeezes will extinguish a combustion-chamber fire. A dry-chem extinguisher will work well, too, but you can't carry one in a coat pocket, and it'll probably create a dry-chemical cloud in the burner room. If the flue pipe is attached, less dry chemical will escape into the room.

Keep Away From the Burner Door and Peephole. Should the burner cycle on and another puffback occur, a sheet of flame could blow out of the peephole or furnace door. At one incident, we were called on the report of an explosion in the basement of a private home. When we arrived, there was thick, hot black smoke pushing out of the basement door under pressure. The storm door was actually pulsating as the pressure of the smoke pushed it open, and it would close again as the pressure was relieved. We delayed venting the windows until a charged line was ready to advance on the basement fire. We extinguished the fire without incident, and an inspection revealed that a resident had stored all of her winter clothes on racks within a foot of the oil burner. The resident had reported an explosion, which had woken her up. Shortly after the explosion, the basement had filled up with the heavy, black smoke. The boiler's fire door was found open and the flue pipe had been blown out of the chimney. This was a classic example of delayed ignition. Improper storage practices resulted in ignition of the clothes, possibly as burning oil was splashed onto them from inside the fire box or as a tongue of flame licked out. Thus, another oil burner emergency escalated into a basement fire.

OTHER HEATING SYSTEM PROBLEMS

Blocked or Disconnected Flue or Chimney

The flue can become dislodged as a result of a puffback, deterioration, or by careless work done on or around the flue. A dislodged or blocked flue is potentially deadly. It will cause the combustion gases, including carbon monoxide, to spill into the boiler room.

There is an accompanying odor related to oil smoke, and it can be considered a safety factor. If CO-laden oil smoke fills a house and the occupant is awake, he'll notice the odor of the burning oil, not the carbon monoxide. If CO-laden fumes from a gas heating unit fill the house, the occupant won't have that warning.

The signs of a blocked or disconnected flue or chimney are smoke that fills the burner room and an oily odor. The residents may complain of flu-like symptoms as a result of the carbon monoxide. If the flue or chimney is blocked or disconnected, shut off the emergency remote, close the fuel valve, and ventilate. Check the residents for signs of carbon monoxide poisoning, and take readings for carbon monoxide. If high levels of CO are found, wear SCBA and remove endangered occupants. Administer oxygen to any victims of CO.

Oil Spills

Anytime liquids are transported, there is a chance of a spill, and fuel oil is no exception. Although its ignition temperature is low enough not to present an immediate threat, oil vapors can make the occupants of a building sick. If the oil becomes contaminated with a low-flash-point liquid, the chance of ignition becomes a hazard.

The appropriate action to take depends on the size of the spill. If the spill is small, use an appropriate absorbent, shovel it up, and remove it. A large spill requires special equipment. This equipment should be available from an oil distributor.

When a building is converted from oil to gas heat, the oil fill pipe is often left in place even though the tank is removed. This sets the stage for a massive spill. If the driver delivers oil to this now-disconnected pipe, he can introduce as much as 275 gallons of oil into the basement. After such an occurrence, you'll be faced with a spill beyond the capability of most fire departments, and you'll need the assistance of the oil supplier. If the delivery truck isn't on the scene, you won't know who the supplier is. Look in the building mailbox to see whether the driver left a delivery ticket. It'll have the name of the company, as well as how much oil was delivered. If the driver realized his mistake, he may have just taken off without leaving a ticket. If you can't identify the company responsible, you'll probably need the assistance of a government agency to effect the cleanup.

The leak should be contained by the building walls, but if it isn't, try to prevent the oil from escaping and entering the sewer system or adjacent buildings. To slow down the generation of noxious fumes, cover the leak with a blanket of foam.

WATER HEATERS

It's estimated that there are about one hundred million residential water heaters in use in the United States. It's also estimated that about one percent, or one million of them, are getting ready to malfunction as a result of improper installation or neglect.

In January 1982, at the Star Elementary School in Spencer, Oklahoma, a water heater malfunctioned. The heater blew up in the cafeteria during lunchtime. Six children and one teacher were killed, and forty-three others were injured. The roof and one wall of the building collapsed. There were two causes behind the explosion: a faulty gas valve and a temperature-pressure relief valve that had been altered.

Gas, oil, and electric water heaters can be found in most buildings, both residential and commercial. Usually they present no danger. A water heater can, however, become a time bomb if it isn't professionally installed, if it malfunctions, or if someone tampers with it. Whether due to age, misuse, or poor maintenance, a water heater can explode, showering metal and glass about the area. Under so much pressure, projectiles from the explosion can penetrate walls and floors. In the Spencer incident, the tank traveled 130 feet, propelled by steam.

Oil-fed water heaters are subject to many of the problems of oil burners, as previously described. Gas heaters are subject to the problems of gas burners, as described in Chapter Three.

The water heater is composed of a metal shell surrounding a water-containment vessel made of glass. It's like a porcelain-covered metal pot. Heated water expands. If it's heated enough, it becomes steam and expands to many times its original volume. The problem is that neither the steel nor the glass expands with it. As pressure builds up in the containment vessel, the expanding water can cause the tank to rupture violently. Safety devices should prevent this scenario from occurring. Unfortunately, safety devices don't always work, and they're subject to tampering.

Water Heater Safety Device

The temperature-pressure valve, located on the top or side of the tank, is designed to vent pressure should it become excessive. Typically, on residential heaters, it'll function if the water temperature reaches 210°F or if the pressure

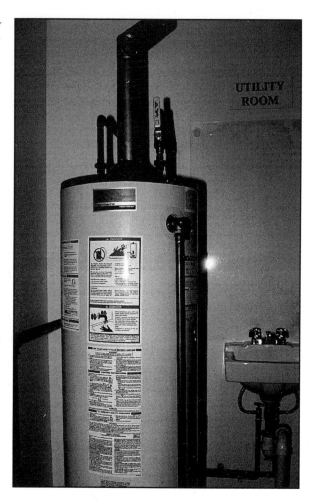

The discharge pipe of a water heater typically runs down the side of the tank to the floor.

exceeds 150 psi. As the temperature reaches 212°F, the water converts to steam and expands 1,700 times. This increases the pressure in the tank and can result in a violent explosion. The temperature-pressure valve is designed to vent the excess pressure by allowing hot water to exit through a drain line before an explosion occurs. The drain line is usually found running down the side of the water heater toward the floor.

If the device is defective, if it has been tampered with, or if it has been clogged with debris from inside the tank, it might not open. If it doesn't open and the water continues to heat up, the tank can fail explosively. A homeowner, seeing water coming out of a properly functioning drain line, might plug the line to prevent water from spilling onto the floor. This creates a potentially deadly bomb that can explode at some later time.

In one incident, fire units were called to a home for a report of an explosion in the basement. On arrival, they found a water heater that had been lifted from its mountings and projected across the room into a bathroom, where it had shattered the toilet bowl. Had the bathroom been occupied at the time, there surely would have been a fatality.

Symptoms of a Defective Temperature-Pressure Valve. Steam or hot water discharging from the temperature-pressure valve; or steam, rather than hot water, discharging from an opened hot water faucet, would both be signs of a defective temperature-pressure valve.

Low-Water Cutoff

Basically, the low-water cutoff is made up of a float attached to an electric relay. When the water level gets too low, the float opens a relay, which then cuts off the heat source.

A water heater should automatically fill up as water is drawn from it for household use. If it doesn't, the low-water cutoff should shut off the heat source if the water drops below a prescribed level. A problem with the electrical relay can prevent water from entering the tank, and the remaining water can become overheated, possibly to the point of steam. If the heat source doesn't shut down, and if the temperature-pressure valve doesn't function, a pressure explosion can occur.

If the tank develops a leak and the water drains out of it, the low-water cutoff should shut off the heat source. If it doesn't, the glass tank and metal shell can become red hot. This will present a possible source of ignition for combustibles that are too close to the tank. Should water be fed to the tank at this point, the resulting steam explosion could prove deadly to anyone in the vicinity.

Symptoms of a Defective Low-Water Cutoff. A cherry red water heater and the smell of hot metal are signs of imminent danger. This condition can occur in a hot water boiler as well. An overheated water heater can become hot enough to ignite nearby combustibles. If the low-water cutoff fails and cool water is pumped into the red-hot tank, a violent steam explosion can result.

Fire department action at both of these emergencies should be to shut down the water heater and fuel supply. Take care not to allow cool water to enter an overheated water heater. You should also avoid contact with the outer shell of the tank. It will be hot enough to cause serious burns. When shutting down the water heater, remember to shut both the electrical power and the fuel supply to the burner.

Safety Shutoff

The safety shutoff stops the delivery of gas to the burner if the pilot flame is extinguished. As long as the pilot light is lit, a thermocouple generates enough electrical energy to activate an electromagnet that holds the gas valve open. If the pilot is extinguished, the current stops flowing and the valve closes. If the valve sticks open or is blocked by debris, gas will continue to flow, and it will leak into the atmosphere.

Delayed Ignition (Gas Water Heaters)

Under normal operation, the gas that flows into the combustion chamber of a gas water heater is immediately ignited by the pilot flame. Blocked burner pots, dirt accumulation, and low gas pressure can allow gas to build up in the combustion chamber. When the gas ignites, a small explosion occurs and a flame will be projected beyond the chamber. The homeowner will usually report having heard a loud whooshing sound prior to noticing the flames near the water heater.

Responding to Water Heater Fires

Determine the Status of the Gas Flow. Since the heater itself is made of metal, it won't contribute fuel to the fire. It can, however, be the cause of a fire. If combustibles are placed too close to the appliance, they can be ignited should a delayed ignition cause a flame to puff out of the combustion chamber. This can happen if the flue is blocked or if there is a continual downdraft. The presence of a draft hood should minimize this possibility. Gas pressure that is too high can cause the appliance's regulator to fail. The result will be flame extending beyond the combustion chamber.

The pilot light can be extinguished by an interruption of the gas supply, a strong downdraft, or the explosion that resulted from a delayed ignition. If the pilot light goes out, safeties should shut off the gas flow to the pilot and burner. If the safety fails, gas will flow, possibly to an ignition source. If the fuel is propane rather than natural gas, the gas will remain in the house rather than go up the flue. Propane, being heavier than air, will hug the floor until it reaches an ignition source.

If the fire is a structure fire, involving nearby combustibles, ordinary tactics will suffice, but you must consider the possibility of a gas-fed fire. If you notice a blue flame, don't extinguish the fire. Treat this as a burning gas leak and protect nearby combustibles. Shut down the gas at the appliance. If that isn't possible, then shut it down at the gas meter. Once the gas has been shut off, continue extinguishment per your normal SOPs.

The main danger is that leaking gas will ignite explosively. If the fire is accompanied by a low-water condition, the tank may be red hot and, if water is added, it may rupture violently.

SUMMARY

Home heating emergencies and fires continue to be a common response. Knowing what type of heating system you're dealing with, what fuel is being used, the signs that an incident might be dangerous or escalating, as well as what action to take, will protect you and your fellow firefighters at these incidents. Understanding how the heating system works, as well as how and why it can malfunction, will reduce your risk of injury even more.

Even though these responses are common, they shouldn't be treated as routine. A whole host of factors can turn a seemingly routine incident into a deadly one, or an apparent emergency into a fire. Knowledge is the key to successfully and safely mitigating the emergency or extinguishing the resultant fire.

STUDY QUESTIONS

1. The most common home-heating fuels are _____ and _____.
2. Based on its viscosity, fuel oil is separated into how many grades?
3. The capacity of a fuel tank in a private dwelling is typically _____ gallons.
4. To ignite the fuel oil in the combustion chamber of an oil burner, the fuel is _____, thus increasing its surface area
5. Safety devices allow an oil burner pump to operate for no more than about _____ seconds if no flame is present.
6. This oil burner emergency occurs when unburned atomized fuel is ignited at the start of the burn cycle, resulting in an explosion.
7. The burning of pooled oil in the combustion chamber after the burner has shut down is known as _____.
8. This condition occurs when the flame repeatedly jumps away from and then returns to the burner nose cone.
9. Since the white ghost is a potentially explosive situation, you should treat it as a _____.
10. On arrival at an oil burner emergency, should you close the furnace door if you find it open?
11. After shutting down the power and the fuel supply valve, what is the proper way to extinguish a fire in the combustion chamber of an oil burner?
12. If smoke is filling the burner room accompanied by an oily odor, suspect _____.
13. The primary threat of an oil spill in a basement is that the vapors will _____.
14. Typically, on residential water heaters, the temperature-pressure valve will function if the water reaches _____ or if the pressure exceeds _____.
15. Name two symptoms of a defective temperature-pressure valve.
16. Name two symptoms of a defective low-water cutoff.
17. When the homeowner reports having heard a loud whooshing sound prior to noticing fire near the gas water heater, what problem is usually indicated?

Chapter Three
Natural Gas Fires and Emergencies

Of all gas fuels used in the United States, natural gas is the most common. More than half of American homes use natural gas for heating and cooking, and it is being installed in 60 percent of all new homes. Natural gas is composed mostly of methane, with varying amounts of ethane, propane, butane, and small amounts of carbon dioxide and nitrogen. The natural gas intended for domestic use contains 70 to 90 percent methane, with ethane making up most of the remainder. When demand is high, usually in the winter, liquid petroleum gas and liquid natural gas may be used to augment the supply. This practice is known as peak shaving.

Natural gas is nontoxic, but it is an asphyxiant and can replace oxygen in a tightly sealed room, possibly causing suffocation. Deadly carbon monoxide, a danger found in manufactured gas, is not present in the natural gas used today.

ODORIZATION

Since it is undetectable without metering equipment, it is a requirement that natural gas be odorized so that a leak can be detected by the user. Typically, mercaptan is the odorant added to natural gas. This odorization enables us to smell gas at levels as low as one-tenth of one percent in air. The addition of odorant is a crucial safety factor because it makes the gas identifiable long before it becomes dangerous. The hazards associated with leaking natural gas are well known, and as a result, any hint of the telltale odor can be quickly reported to the utility company or fire department. Sleeping occupants of a dwelling, however, cannot detect leaking gas because the human sense of smell is dormant as we sleep. Even when we are awake, our sense of smell can be turned off by extended exposure to a substance. Thus, a gas odor might become undetectable after a period of exposure to it. Occasionally the fire department will be called to investigate an odor of gas in the hall of an apartment house. When personnel locate the apartment with the offending appliance, the occupant, desensitized by lengthy exposure, might not even be aware of the gas leak. Besides the effects of lengthy exposure, there are certain individuals who, as a result of a medical condition, have lost their sense of smell. Firefighters, too, in searching for the source of a gas odor can become inured to the scent, and so lose the trail. When searching for the presence of gas or the source of a leak, it is helpful to step outside occasionally to "reset" your olfactory sense, and then to return inside to continue the investigation. Another option is to call in a fresh nose, someone who has not been exposed to the odor. Still, you shouldn't rely on firefighters alone as a means of detecting gas. Properly calibrated gas indicators can sniff out natural gas at levels far lower than the best educated nose, and you should use them at all reported gas leaks. Although sniffing the air can usually alert you to the presence of gas, it cannot conclusively tell you that gas is *not* there, nor can it tell you the degree of hazard posed by a gas-air mixture.

Natural gas is about two-thirds as dense as air. This means that it won't collect in low spaces, whereas propane and butane will. Rather, natural gas will diffuse throughout whatever space it is confined in. Since it is lighter than air, it will easily vent out of a structure. Usually all that is needed to vent a home of natural gas is to open the windows at the top and bottom, as well as the doors. The gas will quickly flow out of the building and dissipate into the outside air. During those times when the utility is using peak shaving methods, the heavier gases propane and butane in the mixture will slow down the venting process.

The flammable range of natural gas in air is 4 to 14 percent. Its ignition temperature can vary between 1,000°F and 1,200°F, depending on how much methane is present. Typically, natural gas is delivered to appliances in the home at $1/4$ psi, or 7 inches of water column.

Liquid petroleum gas, or LPG, is composed of propane, propylene, butane, and butylene. These are all colorless, odorless, tasteless gases that, like natural gas, must be odorized to warn of leaks.

CHARACTERISTICS OF COMMON GAS FUELS

Gas	Vapor Density	Ignition Point	Flammable Range
Natural gas	.55–.65	1,000°F–1,200°F	4%–14%
Propane	1.6	842°F	2.1%–9.5%
Butane	2.0	550°F	1.6%–8.5%

Appliances, designed for use with a specific gas, are marked to indicate what type of gas should be used, but many can be safely converted to work with other gases.

DISTRIBUTION SYSTEMS

Natural gas is delivered to customers via pipeline. It can travel more than 1,800 miles before reaching its destination and be pressurized to as much as

1,000 psi, moving at speeds of as much as 15 mph. It is pushed on this journey by a series of compressor stations that also scrub the gas, removing dust and condensate from it. A rupture or leak in such a line can constitute a major emergency that may be beyond the scope of your department to control. A large gas leak, if ignited, will radiate heat in all directions, threatening nearby structures. Such a leak can be cataclysmic. In March 1994, a rupture occurred in a 36-inch gas transmission line in Edison Township, New Jersey. Within ninety seconds, the escaping high-pressure (975 psi) gas ignited, and the resultant fireball rose five hundred feet into the night sky. The rupture was the result of a previous gouge in the pipe made by excavating equipment sometime between 1986 and the date of the incident. Damage amounted to more than 25 million dollars, as the fire incinerated eight multifamily structures. Extensive mutual aid was required, as well as technical assistance from utility and pipeline personnel.

Natural gas is delivered to homes and businesses from street mains that typically run parallel to the curb. From these street mains, service lines branch off to bring gas to individual occupancies. The mains can contain low- or high-pressure gas. In New York State, gas that is pressurized above 2 psi is considered high pressure. Various textbooks regard pressures above 6 psi as high pressure. Low-pressure gas, often found in older sections of cities, sometimes must have its

The high-pressure regulator is found before the meter on the supply side of the piping.

pressure boosted to adequately supply demand at certain occupancies. High-pressure gas transmission lines can be pressurized to as much as 350 psi and must be reduced to a usable level before entering the customer's appliance. Each end user of high-pressure gas has a gas regulator installed before the meter to lower the pressure down to a usable $1/4$ psi. Individual appliances have built-in regulators integrated into their gas piping. If excess pressure is allowed to reach these built-in regulators, however, they won't hold their set point, and as a result will allow some of the excess pressure into the appliance.

NATURAL-GAS HAZARDS

Air in a Gas Line

This hazard can occur if work is done on a gas line and the line isn't properly bled afterward. Air can also enter a gas line as a result of meter tampering. If air is left in the line, it will travel through the pipe with the flow of gas and eventually exit the line at the appliance. The result can be that an air slug passes through the appliance's burner, extinguishing the flame. An extinguished flame translates into a gas leak. A stovetop burner that was being used to heat a pot will leak gas into the kitchen if an air slug puts out the flame.

The odorant in the escaping gas is easily detected by the occupant, but if no one is home, the gas will continue to build up until a neighbor or passerby notices it and calls the fire department or gas company. If the leak goes unnoticed long enough, if the gas finds a source of ignition, and if the gas is within its explosive range and confined, then an explosion can result. If not sufficiently confined, the gas can ignite in a flash, igniting other combustibles in the area.

Water in a Gas Line

This hazard can occur if a gas appliance is submerged in water. If a basement floods and the water reaches a gas heating unit's burner, it will extinguish the flame. Safeties should keep gas from leaking into the room, but if the burner is reignited before being properly serviced, the safeties can malfunction. If water enters the gas line, it can cause a flameout in the same way that an air slug would.

After a large area suffered flooding, a gas company, in the course of checking out the various gas services in the area, discovered water in the gas line at the level of the second floor. The flooding had occurred in the basement, but the low-pressure gas was able to push the trapped water up to the second floor. Had the gas company not discovered this, the possibility of a flameout would have existed on the second floor and below. In addition, water that finds its way back to the meter can cause it to malfunction.

Although you may not think of calling the gas utility to the scene of a structural flooding incident, it would be prudent to do so if the gas service is involved. While awaiting their response, shut down the service to the building.

Increased Gas Activity

Too much gas supplied to an appliance can cause the flame to rise above the burner. Curtains, cabinets, and other combustibles that, under normal circumstances wouldn't be in danger, might be ignited by the suddenly higher-than-normal flame.

In Brighton, New York, in September 1951, damage to a distribution system gas regulator allowed gas at 30 psi to flow into homes designed for low-pressure gas. The results were devastating. Serious damage or destruction was inflicted on thirty-three homes, and a lesser amount of damage to fourteen more, all within a time frame of an hour and a half. Water heaters were engulfed in flame, flames on stoves shot two feet in the air, meter cases failed, and explosive atmospheres ignited within homes. The Brighton Fire Department received assistance from thirty-six mutual aid companies, and the situation was only brought under control after the gas company shut off the flow.

In McDonald, Ohio, on Thanksgiving Day 1986, a utility company gas regulator malfunctioned, causing a gas surge to seven hundred homes, forty-two of which suffered fire damage, along with two businesses. Some residents reported that the gas surges into their furnaces sounded like freight trains in their basements. Some pilot lights were simply blown out; others torched to as high as five feet. Today, many utility companies have redundant safety systems in place to prevent such an occurrence. Regulators, backup regulators, and automatic shutdown features are meant to prevent the type of incident experienced by the residents of McDonald and Brighton.

Combustion Explosion

If leaking gas is confined to a portion of a building, the right conditions can set off a combustion explosion. Besides confinement, the gas must be within its flammable limits and must encounter an ignition source. Once ignited, the gas-air mixture will quickly burn, and this burning will be accompanied by a rapid buildup of heat in the enclosed area. This heat is absorbed by the contents of the room, the structure, and the air in the room. The heated air expands rapidly, doubling in volume for every 459°F of increase. Most buildings aren't strong enough to withstand such mounting pressures, which can range from as much as 60 to 110 psi. For purposes of comparison, most buildings wouldn't be able to withstand a pressure increase of more than 1 psi without failing.

The failure itself that results from a combustible gas explosion can take the form of anything from blown windows to burst walls and floors. A firefighter standing in a doorway would be blown out of the area and burned by the heated gases and flame. A firefighter standing near the building could be caught in its collapse or struck by flying debris.

Just because a gas is lighter or heavier than air doesn't in and of itself set the stage for a combustion explosion. Typically, air constitutes about 90 percent of all flammable gas-air mixtures. Hence, such mixtures have about the same density as air alone. It isn't necessary for the gas to fill the entire space for an explosion to occur. It can collect in only a portion of the enclosure and still result in a combustion explosion, given that it finds a source of ignition.

Loss of Odor

Odor isn't a definite indicator that natural gas is present. Odorants can be scrubbed out of the gas if it passes through a filtering material such as sand or dirt. To some extent, odorants can even be absorbed by iron gas pipe. This occurs only in new iron gas piping, and it will continue until the pipe is saturated with mercaptan. When utilities install iron piping, they test the gas delivered to the customers to ensure that the odorant hasn't been absorbed along the way. The testing continues until the saturation point is reached and the metal stops absorbing the odorant.

Scrubbing the odorant out of natural gas can render the gas undetectable by smell, allowing dangerous amounts of it to accumulate if a leak is present. The only way to detect scrubbed gas is with a flammable gas indicator, which can also be used to locate the source of a leak. A digital flammable gas detection instrument is a reliable indicator of the relative hazard. It enables you to determine not only that natural gas is there, but also its concentration and whether or not it is within, above, or below its flammable range. Most such indicators have an alarm that can be set to sound when the concentration approaches the lower flammable limit.

Unfortunately, many firefighters still rely on their sense of smell to detect and locate a natural gas leak. At one incident, a woman complained to firefighters of a recurring gas odor in her fourth-floor apartment. The members sniffed around her apartment and the public hall but found no trace of gas. A combustible gas indicator, brought into the apartment, detected gas where the firefighters could not. Still unable to locate the source of the leak, however, the members expanded their search to the apartment on the floor below. That apartment, too, was found to contain small amounts of gas that were undetectable without the device. The source of the gas was finally located in a first-floor apartment, where a cigarette-smoking occupant swore that there was no leak. Firefighters entering the apartment, however, noticed a very strong smell. The leak was found in the flexible tubing connecting the stove to the gas pipe. The escaping gas was entering a hole in the wall and rising into the apartments above. The lucky occupant didn't notice the smell of the gas and was unaware of the danger posed by his lighted cigarette.

The indicator used at this incident had no digital readout, but instead clicked more quickly, Geiger counter-style, in the presence of greater concentrations of natural gas. Without it, the firefighters would have walked away and taken no action. I wonder how many gas leaks have been missed by members who, not smelling anything, have determined that there was no leak.

A gas detection instrument is an invaluable tool when investigating gas leaks. If the fire company at the scene of a suspected leak doesn't have such a device, the officer should call a company that does. Your utility company will gladly supply the gas indicator and trained personnel to assist in the search for a suspected leak. In fact, the utility company might want to be called to all such operations, since it is in their best interest to avert danger. Certainly they should at least respond to all confirmed gas leaks, even minor ones.

Regulator Failure

A pressure regulator can fail if debris enters it and prevents the valve from seating properly, or if its diaphragm fails. High-pressure systems have a vent pipe as well as a regulator. If the regulator fails, much of the excess gas delivered will be diverted through the vent pipe and to the outer air instead of into the structure.

Because a particular type of wasp was fond of setting up house in gas vent pipes, utility companies placed protective screens on the pipes. On older vents, a coiled spring was used to keep the vent pipe free from clogging with debris or bugs. When gas flowed out of the vent, the coiled spring expanded and cleared

If the regulator fails, high-pressure gas will be discharged through this vent.

the opening of any accumulated debris. This type of vent has been replaced with a simple mesh screen that serves the same purpose but that has no moving parts. In the event of a regulator failure, even a properly operating vent will allow some of the excess gas into the building's piping system and subsequently to the various gas appliances. At these now overpressurized appliances, the result of excess pressure will be felt. The gas flame can either be blown out, possibly resulting in a gas leak, or the flame may grow, possibly to dangerous heights. If the vent is clogged, then even higher pressures will be sent through the meter to the appliances, creating an even greater hazard. Some appliances have their own pressure regulators, but these, too, can be overcome by the excess gas flow.

The failure of a regulator will be accompanied by a strong smell of gas and a hissing sound in the area of the vent. In the case of a regulator failure, the fire department will have to contact the utility because the regulator will either have to be repaired or replaced. To shut off the gas supply to the regulator, shut the quarter-turn valve located before the regulator. This will be easy enough if the regulator and meter are located outside of the building, since the gas will dissipate into the atmosphere. We must, however, remain alert to possible sources of ignition. A fog stream can be used to disperse the gas and protect the firefighter shutting off the gas. If the regulator and meter are inside the building, you must enter with caution, locate the meter, and shut off the gas before the regulator. You must then search and vent the entire building. The failed regulator may have caused a fire inside the building if the escaping gas was ignited, or substantial amounts of gas may have accumulated, resulting in an explosion hazard. Although it is unlikely that the gas will be present in such quantities as to place the occupants in danger of asphyxiation, this is a possibility, and a search must be conducted for overcome occupants. At the same time, you should identify and remove potential sources of ignition.

If the meter and regulator are located inside the building, it may be prudent to shut off the gas at the curb valve, if you can locate it. If one is present, it should be on the sidewalk under a square or round metal box with a metal cover. The size of the box will vary with different utility companies. The cover must be removed and the valve turned with a special wrench, which in most cases will be one-quarter turn to curtail the service. Older systems may require multiple turns to close the valve. You should know which type of valve you will encounter in your district. Once the gas supply has been shut down, you can search and vent the building without the danger of an active interior gas leak.

Delayed Ignition

The main hazard of gas-fired burners is that of delayed ignition. Under normal operation, the gas that flows into the combustion chamber is immediately ignited by the pilot flame. Blocked burner ports, accumulated dirt, and low gas pressure can allow gas to build up in the chamber. When the gas ignites, a small explosion occurs, projecting a flame beyond the combustion chamber, possibly igniting nearby combustibles. The homeowner will usually report having heard a loud whooshing sound prior to noticing the fire near the water heater, furnace, or boiler.

FLAMMABLE-GAS FIRES

Essentially, a flammable-gas fire is a combustion explosion that never occurred, either because it was unconfined or because it was ignited too quickly.

I recall responding to a reported gas main break on a busy street. On arrival, we were greeted with ten-foot flames rising out of a slit in the street. The slit had been made by a large diamond-tipped rotary saw used by workers to cut asphalt. The operator said that he had been cutting the pavement for the cable TV company, which had been installing cables underground. The operator explained that all had gone well until suddenly he heard a loud hissing sound accompanied by a strong smell of natural gas. Realizing the danger, he stopped working and removed his cutting machine from the area. As he ran away, the escaping gas ignited. He was lucky to have escaped injury.

A cable company employee was cutting a trench in the street when he inadvertently severed a gas line. He escaped before the gas ignited.

76 • Responding to "Routine" Emergencies

Firefighters operating a fog stream to protect utility workers as they cut and cap a leaking gas main. The fog stream will prevent the flames from igniting vapors released by the cutting operation.

The narrow cut had been made in the middle of a four-lane street, and there were no structural exposures. After notifying the gas company to respond, we set up fire lines, keeping everyone a safe distance away. Electrical wires hung directly over the flame. The insulation on the wires was burning, so we notified the electric company to respond. Luckily, the wires didn't fail.

Realizing the explosive potential of leaking gas, we didn't extinguish the flame. Instead, we stretched precautionary lines and waited at a safe distance. The gas company's emergency crew dug through the pavement some distance away on both sides of the rupture. They intended to cut the line and cap it at both locations. We were requested to position protective lines between the flames and the workers. The idea was to prevent gas released through the twin cuts from reaching the flames, yet we had to do so without flooding any of the three openings. We used fog lines on both nozzles to direct gas fumes away from the flame. The gas lines were cut and capped without incident.

Tactics for Flammable-Gas Fires

Shut Off the Gas Before Extinguishment. Don't extinguish a gas fire unless the escaping gas can be shut down. Extinguishing a gas fire may allow escaping gas to accumulate and, if an ignition source is present, to reignite. If the gas is confined, the result may be a combustion explosion. Safety dictates that a gas fire should not be extinguished, but rather, that the gas supply be turned off.

Protect Exposures. If a gas fire is exposing combustibles, protect them with a hoseline. A narrow fog stream works well for this purpose. Be careful not to extinguish the gas-fed flames. Extinguish the burning combustibles if doing so won't extinguish the gas flame. If the flame is impinging on a building, operate a fog stream on the exposure. If the burning gas threatens surrounding buildings, you may need to position multiple precautionary lines.

I once responded to a cellar fire in a vacant apartment building. The gas meters and much of the piping, including the shutoffs, had been removed from the building, and the escaping gas ignited. The flame had caused the plaster to fall, had ignited the ceiling lath, and was threatening to extend to the floor above. A line was stretched to the meter room but wasn't operated for fear of extinguishing the gas flame. Another line was stretched to the floor above, where firefighters were searching for extension. We called the gas company for help.

After consulting with the utility company's emergency crew, it was decided that we would extinguish the flame and the repair crew would cut and cap the gas line in an adjoining room. After putting out the flame, we kept a fog line opened in the room to prevent reignition. We also positioned a precautionary hoseline to protect the personnel cutting the gas line. The line was cut and capped without incident. This incident occurred more than twenty years ago. Today, as a result of improved technology, the gas company might instead choose to inject grease or expanding foam into the line, stopping the flow of gas without the danger associated with cutting and capping a gas-filled pipe.

Dissipate Any Leaking Gas. If you extinguish a gas flame, you'll be left with a gas leak. If you inadvertently extinguish the flame, use a wider fog spray to dissipate the gas so that it won't ignite. If the leak is indoors, ventilate the area, remove all ignition sources, and try to direct the escaping gas away from any remaining sources of ignition.

INDOOR GAS LEAKS

Never underestimate the potential of a natural gas leak. A seemingly small, routine leak can become deadly if the leaking gas accumulates somewhere in the structure and finds a source of ignition. It is important that we do not provide that source with our radios and hand lights. The simple act of turning on a hand light or a wall lamp, or even walking over carpeting and generating static, could trigger an explosion. The ringing of a telephone or doorbell could also ignite the gas.

Tactics for Indoor Gas Leaks

Position Apparatus Safely. Too often we take gas leaks for granted and don't consider the safe placement of apparatus. Although most of us will never encounter a gas leak that results in an explosion and the collapse of a building, your SOPs should require that apparatus be positioned out of harm's way. At the very least, this means that it should be parked out of the possible collapse zone of the building. It should also be out of the path of flying debris in case the leak ignites explosively. Have the engine take a position at a hydrant or other water source that is in a safe location. The ladder apparatus should take a position that will allow it access to the building frontage but that doesn't place it directly in front of the building.

Determine Whether the Report of a Gas Leak Is Accurate. It may be that the occupant has smelled some other odor and mistaken it for gas. I have been called to buildings for an odor of gas only to find someone burning incense or using paint in another apartment. An experienced firefighter can usually identify the type of odor present, but he must be wary of odors that might be masking gas. Remember, the only sure way is to test the air with a gas indicator.

Occasionally it turns out that there is no gas leak but that a landlord dispute is going on and that one party is trying to harass the other by bringing in the fire department. Recently I was called to a two-family home by the landlord, who was reporting a gas leak in the downstairs apartment. The apartment was locked and the tenant was out. The landlord wanted us to force the door. He also mentioned that he wanted to evict the tenant for long-term nonpayment of rent. Since there was no apparent odor, we suspected that the landlord had called us simply to gain access to the apartment so that he could remove the occupant's belongings. Instead of forcing the door, we took readings at the door and windows of the apartment, but our gas indicator showed no detectable level of gas. We decided not to force the door. If you are in doubt, you must force entry and search the apartment, but you should consider what you will do when, as in the above example, you doubt the validity of a call for a gas leak. What exactly is your liability if you force entry into the apartment, and what is your legal exposure if you don't? This is a dilemma that you must consider before you are actually faced with it on a call.

A suspected natural gas leak might also turn out to be the result of sewer gas seeping into a home from an open trap on the home's sewer line. Flammable and odorous liquids poured down the home's drains can seep out of an unsealed trap. It is also possible for vapors from liquids that have been poured into outside sewer lines to find their way into a home through that same unsealed trap. The solution is to seal the trap, which isn't always possible, and then to flush the building's sewer system by running water in the sink and flushing the toilet several times. Sealing the trap will prevent the odor from entering the home, and running water will move the liquid and vapors out of the home's piping. Ventilating the home should remove the residual odor. In one such instance, the odor went from one house to the next as we flushed the drains of affected homes. Finally, we opened several hydrants and flushed the sewers. This eventually solved the problem, or at least moved the source of the odor through the street sewer lines and out of the area. In such instances, it is important to monitor the air in the home and perhaps the sewers. If a possible flammable gas-air mixture is present in the building, ventilate. Evacuate the building, if necessary. Even an odorous, nonexplosive mixture of a flammable liquid can make the occupants ill and may also require ventilation and evacuation.

Try to Determine When a Gas Odor Was First Noticed. A sudden strong odor could indicate a serious leak. A slight odor that comes and goes could be from an extinguished pilot light and will *probably* be less serious. On the other hand, the slight odor could be seeping into the building or apartment through a common wall. Your search should not be limited to one occupancy, for you can never be sure that the source you find is the only source of leaking gas in the building. This is where the gas company can help. Utility personnel have the training and the time to make a thorough search for possible sources of leaking gas, plus the skills and equipment to fix them. The Brooklyn Union Gas Company was called to a building for an odor of gas and found seven leaking meters, which they repaired. Had the fire department been the only agency on the scene, it's doubtful that all of the sources would have been discovered. Calling the utility company makes good sense.

Find Out What Type of Gas Is Involved. You must determine whether the call is for a natural gas leak or a propane gas leak. The hazards aren't the same. Since propane collects in low places and doesn't vent as easily, it is important to determine early which gas you're dealing with.

Locate the Gas Leak. The only reliable method of locating a leak is to use a properly calibrated combustible gas indicator. This instrument must be intrinsically safe so as not to pose an explosion hazard. Such gas indicators can measure what percent of the lower explosive limit of the gas is in the air or what percent of the sampled air is natural gas. Since natural gas is explosive between 4 percent and 14 percent, a reading of 1 percent indicates that one-fourth of the explosive amount is present in a given volume of air. While your nose can usually tell you that gas is present, it won't warn you how dangerous the mixture might be. A gas indicator will both help you locate the leak and determine the existing concentration of gas. Utility personnel investigating gas leaks do so with the appropriate indicator, and so should the fire department. A gas indicator that doesn't have a digital readout will be able to detect and locate leaking gas, but it won't be able to determine the level of hazard.

If you are called to a reported odor of gas in a multiple dwelling and there is a slight odor in the hallway, you may have to force entry into many apartments so as to locate the source of the leak. A slight odor in the hall may indicate a more serious leak in an apartment. A gas indicator with remote sampling capability could be used to sample the atmosphere in each apartment by checking at the top crack of the doors. Although you could slip the meter's probe under the door, the

This gas meter indicates the presence of a combustible gas by clicking like a Geiger counter. It lacks a digital display and won't sound an alarm if the gas reaches its explosive range.

probe wouldn't detect the presence of gas if a slight leak only allowed gas to accumulate near the ceiling. Taking readings near the floor isn't a good way to get accurate results. If the door has a security peephole, you can remove it and insert the probe there. Investigating this way can reduce the number of doors that you have to force, resulting in a quicker, less damaging operation. If you still can't locate the source of the gas, you may have to force doors after all. Also consider that the gas odor may be coming from an exterior source, entering the structure through an open window.

The building's gas meters can be located either outside of the building or in its cellar or basement. A meter that registers heavy gas use may indicate which apartment has the leak. This may not be a reliable method if it is dinnertime and many occupants are using their stoves, but in the middle of the night, it may be a good indicator. In rare instances, gas meters are located in individual apartments.

Once you've narrowed down your search, apply a soapy solution onto a suspected leak in a pipe or joint to help you pinpoint the trouble spot. If a leak is present, the solution will bubble. A common source of leaks is the coupling or the flexible hose that connects an appliance to the fixed gas piping. Older flexible connectors were made of uncoated brass and are prone to deterioration. Newer connectors are made of coated brass or stainless steel and won't deteriorate. Moving the appliance to check the connector can actually cause the older type to leak. You should warn homeowners to have older connectors replaced by the newer type. The Consumer Product Safety Commission reports that thirty-eight deaths and sixty-three injuries are attributable to the older-type connector, and the agency has issued a warning. The CPSC recommends that, when gas appliances are checked, the connector be checked with a gas indicator. The older type of connector is recognizable by its brass color. The epoxy-coated flexible connector is grey or black in color. The newer stainless steel connectors are easy to recognize by their metallic shine.

Call the Gas Company. The gas companies that I work with all want to be called to all reported gas leaks. This is good policy. You should, however, be able to control a minor leak without the assistance of the gas company, simply by shutting off the offending appliance. You must also make sure that it is the only source of leaking gas in the home. If you cannot do this, then you need the assistance of the utility. A major leak will always require the assistance of utility company personnel, since they have the tools and training. In any case, you must notify the gas utility when you shut the gas supply to an appliance or to the whole building. They will be required to relight pilot lights before an appliance or full service is turned back on, as well as to test for and fix any leak in the distribution system piping up to and including the meter. The occupant is usually responsible for fixing leaks beyond the meter.

Shut Off the Gas. Whether the incident is a leak or a fire involving gas, you should stop the flow of gas by the nearest available shutoff. Always try to isolate the leak by shutting down only the affected line or appliance. This won't always be possible. Fire in the vicinity of the nearest valve may prevent you from approaching it, or the valve itself might have been damaged by the fire and been rendered inoperable. An explosive atmosphere within the building, accompanied by ignition sources, may mandate shutting off the gas at the exterior. In a collapse situation, it may not be possible to reach the location of the leak, so again, an exterior shutoff may be your best bet.

The order of priority for gas shutoffs is as follows:

(1) *Appliance Quarter-Turn Shutoff.* This valve is located where the appliance attaches to the gas piping. Appliances such as stoves and clothes dryers are usually connected to the home's gas piping by a flexible metallic hose. The shutoff will be at the point where the hose attaches to the fixed piping. Hot water heaters and furnaces are typically attached directly to the service piping. All have a quarter-turn shutoff. It's easy to close this valve by gripping it with pliers, vise grips, or the fork of a halligan. Turn it one-quarter turn, usually clockwise. When the rectangular nut is parallel to the gas piping, it is in the on position and gas is flowing through it. A quarter-turn of the nut, turning it perpendicular to the flow, will shut off the gas. If the valve won't turn, don't force it. Go instead to the next available shutoff on the service side of the meter.

When shutting off the gas to an apartment, make sure that you select the right meter. The quarter-turn wingcock is located on the riser to the left of each of these meters. Unit 1A has been turned off; 2A is still being served.

(2) *Meter Quarter-Turn Wingcock.* This shutoff will be found on the supply side of the meter. If the leak is before the flexible hose, or if the appliance shutoff doesn't stop the flow of gas or isn't accessible, you must then go to a more remote shutoff. In most cases, this will be at the meter. In private homes, the meter will most likely be located where the gas enters the building. It may be found inside the building or on the outside, on the exterior wall. In multiple dwellings, there may be multiple meters, one for each occupancy. They may be located inside the building, where the gas pipe enters the house; in each apartment; or clustered together outside of the building. The most common type you will encounter here will again be the quarter-turn valve. If the building has a basement, this valve will be where the gas piping enters the building wall. If the building is built on a slab, the valve will usually be outside, where the gas piping comes up from the ground.

(3) *Curb Valve.* The curb valve can be found on the house side of the curb if high-pressure gas is supplied to the building, and in some cases in houses supplied by low-pressure gas. In my area, curb valves are found beneath a metal cover, nine inches square, with the gas company's logo on it, the word *gas,* or simply the letter *G.* You must remove the metal cover to expose the quarter-turn shutoff inside. Closing the valve requires a special wrench. The curb valve may

This curb valve cover is marked with the letter G.

not be present in all cases, or it may have been covered by cement work or landscaping done by the homeowner. In some areas, such as New York City, the curb valve may be on the sidewalk across the street from the building being supplied. Known as long service, an arrow on the cover will point to the building that the valve supplies. Your utility company can inform you of any such irregularities in your area.

(4) *Street Valve.* Also known as a service valve, these are typically located on the street side of the curb. In some areas, the street valve can be found on the sidewalk, and in some places where there is no curb, the valve may be found hidden in the grass. In any case, you should never operate a street valve. Your gas utility will be able to locate the valve, identify it, and close it. Such a valve controls large areas of gas supply, and closing it might disrupt service to a substantial number of gas customers. Only the gas company should shut this valve. Closing the valve may in fact not even stop the flow of gas to a particular building. Some distribution systems are fed from multiple directions, and shutting one valve may not curtail the supply. Again, the utility company will know the correct combination of valves to cut off the gas supply to a particular building.

Never turn on a gas valve, gas service, or an appliance once it has been cut off. To do so could allow gas to leak into the building from damaged piping or an unlit pilot light. The gas service should only be restored by the utility company. Before turning the gas service or appliance on, they will conduct a test of the piping and light the pilot.

84 • Responding to "Routine" Emergencies

A cutaway of a street valve, also known as a service valve.

At a major gas leak in the piping of a high-rise housing project, we turned off the service to the entire building. Turning it back on was a monumental task for the utility company. They had to gain access to each of more than one hundred apartments, as each stove's pilot light had to be relit and the piping tested. While you shouldn't hesitate to cut off the gas service if safety concerns mandate doing so, you shouldn't take such action indiscriminately. If possible, consult first with utility company personnel.

Search the Area for Trapped Occupants. If the leak is a major one, or if the gas has leaked for a prolonged period, substantial quantities may have accumulated. If the percent of oxygen in the air is reduced from the normal 21 percent down to 17 percent, motor coordination will be affected. At 10 to 14 percent oxygen in air, a person's judgment becomes faulty and fatigue sets in. At 6 to 10 percent, a person will lose consciousness and die if not removed to fresh air quickly. Although possible, it isn't likely that an occupant will be asphyxiated as a result of leaking gas. Therefore, such a victim will almost certainly be in need of rescue and medical attention. The larger the gas leak and the tighter the building is sealed, the greater the chance that asphyxiation will occur.

The manufactured gas once supplied to homes by utility companies contained deadly carbon monoxide and could be used as a means of suicide. Utilities now deliver natural gas, which contains no CO and is nontoxic.

Although natural gas is an asphyxiant, deadly concentrations of it are rarely encountered at leaks. Anyone, however, who suffers from a heart or lung ailment and is normally oxygen deficient might have an adverse reaction even to slightly reduced levels of O_2. Victims should be removed to fresh air and given appropriate medical treatment.

Our first duty is to save lives, and we must always consider the possibility that someone in the building might be in danger. A quick search of the occupancy to locate any such victims may be called for, but it may also be too dangerous to attempt. You must perform a risk-to-benefit analysis prior to taking such action. If a hazard of ignition or explosion exists, expose as few firefighters as possible to the danger. Interviewing occupants can help determine whether anyone else is in the building and where they might be located. Knowing where to look for victims will reduce the time that firefighters are exposed to danger.

If the concentrations are approaching or over the flammable limits, exercise extreme caution. An atmosphere that is near the lower explosive limits at the door may in fact be within the explosive limits farther into the building. If the building contains a gas atmosphere that is above the explosive limits, opening a door or window will admit oxygen into the immediate area, bringing the mixture into the explosive range. Stretch a precautionary hoseline to protect the searchers. Full protective gear and SCBA will be required because of both the oxygen-deficient atmosphere and the chance of ignition. You needn't stretch a line for a small gas leak that isn't near the explosive range. Firefighters should, however, be ready to deploy one should conditions change for the worse.

Prevent Ignition. Do not bring an ignition source into the gas. Are your radio and hand light intrinsically safe? If not, keying your radio could supply an ignition source. Since switching on your light could trigger a devastating explosion, you should switch your light on before you enter the area, and don't turn it off until after you leave. Don't operate any electrical switches in the area, since they too can create a deadly spark. Beware of carpets and of touching metal objects. Do not remove power from the building by pulling the meter. Gas can travel up the electrical conduit into the meter, and the spark created as you pull the meter could ignite the gas. If you must remove power from the building, do so by cutting the electrical supply lines at a location beyond the explosive atmosphere. While you shouldn't operate any electrical switches inside, you should extinguish pilot lights and any other open flames. Keep in mind that there are ignition sources beyond your control. As you open the door to an occupancy, a cold breeze might cause the thermostat to call for heat, thereby supplying an ignition source.

Ventilate the Affected Area. You can do this as you search for victims or before you enter the building. At most leaks, breaking glass should not be necessary, and natural ventilation should do the trick. Simply opening the windows, top and bottom, will in most cases adequately vent the area. A typical residential structure should vent completely within minutes. When possible, vent a multistory building at its highest level first, then the lower levels. If you do the reverse, the lower levels will be inundated with oxygen as you reach the upper floors. Those lower floors may be entering the explosive range, and you will have to travel back down through them to exit the building. Venting the upper floors first puts you out of harm's way as you drop down to the next unvented floor. This isn't always possible, however, and in most cases, each floor will vent quickly once you open the windows. If a given area is too hazardous to enter, venting from the exterior is an option. It may cause damage to the windows, but not as much as a combustion explosion, and it will be far safer than entering the building to vent.

If the concentrations are heavy, residual gas can accumulate and remain in closets, cabinets, attics, walls, and other enclosed spaces. Such pockets can remain in these areas even after the building has been fully ventilated and gas indicators show it to be safe. If an ignition source is present in any of these enclosures, ignition can occur. Prior to declaring the incident closed, retest these areas and vent them, if necessary.

If you determine that additional ventilation is necessary, consider mechanical means. Since pulling leaking gas through an electric fan (even one that is supposedly safe) might precipitate a disaster, consider using PPV to force the gas out of the building. Take care, however, where you force the gas. Are you pushing it toward an ignition source or into a concealed space? Is it being pushed into an elevator shaft? Are you making the situation more hazardous by forcing the gas into other areas of the building? A safer method is to use a fog stream, either to pull or push the gas out of a window or door.

In most cases, simple natural ventilation will work, and it is the safest, easiest option. If because of atmospheric conditions the gas doesn't vent, open doors and windows on as many sides of the building as possible. Cross-ventilation should speed up the process.

OUTDOOR GAS LEAKS

In January 1967, the Fire Department of New York was challenged by a street-main gas leak of monumental proportions. It required thirteen alarms and a total of sixty-three companies to control this incident. The culprit was a drip pot, which is used to scrub moisture from the pipeline. Its cover had come loose and become partly dislodged.

The escaping gas sounded like a jet engine. Responding apparatus were stalled as they entered the gas-rich atmosphere, and a pumper and a ladder truck were engulfed in flame when the gas later ignited. The radiant heat from the burning gas destroyed a paint factory, a garage, and nine homes, and the fire department was faced with a need to evacuate several densely populated residential blocks. Eight other buildings were damaged by fire. The heat was so intense that tables and doors from the buildings were used as heat shields to protect rescue workers who were evacuating the exposed buildings. In the wake of this incident, various regulatory agencies, as well as the involved utility, determined the cause and undertook the necessary steps to prevent it from recurring. In response, the utility removed the drip pots where possible, and shored and securely fastened the covers on those that remained.

Tactics at Outdoor Gas Leaks

Determine Whether a Gas Leak Exists. As with inside leaks, the reported odor of natural gas may be the result of something else. In my district, there is a sewage treatment plant. When the winds are just right, an odor of sewer gas can settle over a large area. As a result, we are called to reports of multiple gas leaks.

When an odor covers a large area, it can be hard to track down. You will need gas detection instruments to determine whether the odor, whatever the source, is accompanied by a flammable atmosphere.

Gas leaking up from under the ground for a long time will turn the foliage brown and eventually kill it. A heavy underground gas leak located in a field or wooded area can sometimes be found by looking for dead grass, bushes, and trees in the vicinity of the leak.

Call the Utility Company. Outdoor gas leaks can be difficult to locate. Also, supply lines deliver gas at high pressures, and a leak in one can allow large amounts to escape. Patching or plugging them requires special tools and skills. Even if you have firefighters trained to do the job, ruptured or damaged distribution pipelines are a job for utility personnel.

If, due to location, the response time of the utility will be long, and if your firefighters are trained and have the tools, they can attempt to temporarily plug a low-pressure pipe to stop the flow. A low-pressure pipe, depending on its size, can be plugged by applying hand pressure against the leak. Duct sealant, duct tape, and even a wad of paper can be used to temporarily stop or slow a leak in a low-pressure pipe. You can also use rags or a wooden plug in a pipe that has been completely severed. You may have to tape them in place with duct tape. If you do plug the leak yourself, you must notify the utility personnel when they arrive. These temporary measures won't work with a leak in high-pressure gas piping. Such leaks have to be plugged by the gas company.

Beware of the Static Hazard of Plastic Pipe. Supply pipe is made of either metal or plastic. The metal pipe is conductive, and gas leaking from a metal pipe does not pose a static electricity hazard. Plastic, on the other hand, is nonconductive and will allow a static charge to build up as the gas flows freely from the break. This static charge can be as high as 14,000 volts. If a firefighter approaches or touches the plastic pipe, the resulting static spark can ignite the leaking gas. Plastic pipe is so prone to static buildup that air flowing through pipe stacked on a moving utility vehicle delivering plastic piping was able to generate enough of a charge to give the truck driver a shock as he later unloaded the pipe.

To minimize the risk of a spark being discharged by a leaking gas pipe, it is recommended that a rag, wet with a soapy solution, be placed on the pipe and in contact with the ground, or that soapy water be sprayed onto the pipe. Soapy water will break the plastic's surface tension, whereas plain water will simply bead up, making an insufficient path to ground. Keep the pipe and cloth wetted with soapy water, since, if they dry, the flow of electrons to ground will be broken. Do not, however, apply so much water that the ditch fills up. The utility workers may have to enter the hole to stop the flow of gas.

Although a small plastic pipe can be folded over on itself to stop the flow of gas temporarily, you must consider static electricity as a possible source of ignition, and so take precautions. Utility workers must receive special training before being allowed to work with plastic piping. If your firefighters haven't been trained specifically to handle plastic pipes, leave this seemingly simple task to the repair crew.

Approach the Leak From the Upwind Side. Units responding to a report of an outdoor gas leak should consider approaching the site with the wind at their backs and not bringing their apparatus too close to the source. If there is a strong odor of gas, it might be better to walk a block to the scene than to suffer the same fate that the New York City pumper and ladder truck did in the incident described above. Consider also that the wind may shift, perhaps threatening parked apparatus that had been thought safe. Shut down apparatus and electrical systems unless they are needed for the operation.

Determine the Scope of the Incident. Is the leak major or minor? Will you be able to handle it with the resources on the scene, or will you need specialized assistance? It may be the result of an impact delivered by the demolition of a nearby building or a water leak that undermined a cast-iron pipe. Backhoes are responsible for many gas line mishaps, severing them or pulling them right out of the home piping. The latter case creates a leak near the home that will likely result in a buildup of gas in the home. It is a good idea to have contractors register with the municipality before doing any excavating. "Call before you dig" is a national program set up to help contractors locate gas lines and other underground facilities before they break ground.

As mentioned above, gas leaking outdoors can find its way into a building. At the scene of one exterior leak, the utility's emergency crew was concerned that a heavy cover of trees in the area wasn't allowing the gas to dissipate into the atmosphere. They worried that if someone were to light a nearby fireplace or barbeque, the gas might ignite, so they warned all of the nearby residents to refrain from doing so.

Natural gas escaping from a broken pipe under the street can find its way into a building, too. Although gas naturally percolates up and out of the ground, it can be trapped by pavement or even by a layer of frost. Gas trapped in this fashion can travel great distances before escaping into the atmosphere or entering a building. Sand, often used as fill when pipes, conduits, and sewers are buried, is more porous than dirt and allows the leaking gas to seep along sand trails hundreds of feet long into a building. As a result, a seemingly harmless outdoor gas leak can cause a dangerous situation in nearby buildings. Utility companies routinely insert a probe into the ground to detect the presence and migration of gas at such locales. Buildings in the vicinity of an outdoor gas leak must be checked for accumulated gas. Combustible gas indicators are a must for this type of investigation, since the mercaptan may have been scrubbed out of the gas as it passed through the soil.

While responding to a report of a ruptured gas main, firefighters detected a strong odor of gas in the street. The fire department units arrived just before the utility company's emergency response truck. Previous work being done in the street had apparently damaged a gas pipe. Both the fire and utility personnel investigated and found no dangerous condition. With the utility company on the scene and no apparent hazard, the fire department left. Later in the day, firefighters were called back for a reported explosion. Undetected gas had migrated under the pavement and into a nearby brick home, where it ignited explosively. The foundation wall was blown outward, and the building's gas line, damaged by the explosion, was leaking. Gas from the now-leaking pipe ignited, and in turn ignited surrounding combustibles. The stability of the building was in question, and gas was still leaking in from the street. The responding truck company made a quick search of the damaged building. Finding no one home, they exited. The engine company stretched a line and stood fast. Once the gas to the building was shut down by the utility workers, and the building's stability was reevaluated and deemed safe, the engine company entered and extinguished the fire. No one had been home at the time of the incident, and luckily no one had been injured in the blast and the ensuing fire.

At another incident, a gas leak in the street was seeping into a home. The service line had been pulled out of the building's piping by a careless backhoe operator, and gas was leaking in from the gap in the piping. Utility workers entered the building, looking for the location that the gas was entering, but they couldn't find it, nor could they find the meter. Eventually they removed the wood paneling from the cellar wall and found that the gas meter had been covered and that gas was seeping in near the meter, through the wall. When you search for the source of a gas leak, consider that it may be hidden behind the work of a home handyman.

Define the Hazard Area and Protect the Public. Using both your sense of smell and gas detection equipment, identify the hazard area and set up barriers to exclude the public from the danger zone. Stretch lines from a safe location to control any fire that results from ignition of the leak. Initial operations may include evacuation of all in the hazard area. This may range from a fairly simple operation in a rural area to a very involved one in a city. It may require closing streets to vehicular traffic and rerouting trains. The police department can help by controlling crowds and traffic. Set up barriers or use fire zone tape to indicate the danger zone. Call for help early. Consider the wind direction when deciding what areas to evacuate and what areas to rope off from the public.

Determine Where the Leak Is. Is the leak inside or outside? If it's outside, is it seeping into any structure? Is there an immediate explosion hazard? The officer in command must answer these questions early in the operation to respond adequately to the threat posed by the leaking gas.

Determine What Resources Are Needed. Does the incident commander have the resources to control the situation, or will he need help? What help is available? How long will it take for help to arrive? What special tools are required? In all cases, call for help early. Try to get an estimate of how long the utility company will take to respond. If the condition is escalating and the utility's response time is long, it may be necessary to call additional fire department resources to the scene.

Give an Accurate Description to the Utility Company. A precise description of conditions will assist the utility in dispatching the appropriate resources. In a period of high incidence of utility emergencies, it will also allow the utility to prioritize its responses. Knowing an exact location of the leak will enable the utility to locate in advance the appropriate valves for controlling the situation.

Keep Emergency Personnel Out of the Danger Area. Although it may be necessary to enter the potentially explosive area to search for and remove victims, once this has been accomplished, firefighters should either shut down the gas supply or, if they cannot, stretch precautionary lines and stage in a safe area to await the utility company.

Stretch Hoselines to Dissipate the Gas. If life or property is at risk, consider using fog lines to dissipate the gas cloud. Doing so can benefit search operations. In one instance, a leak in a high-pressure gas line located between two closely spaced buildings threatened to allow gas to enter the buildings via their overhanging eaves. If this had occurred, the incident would have gone from an outside gas leak to an interior accumulation of gas. To prevent this, we operated a fog handline into the leak to disperse the gas.

SUMMARY

Fire departments have and will continue to respond to the smallest as well as the largest of gas leaks. When we arrive, often we will be the first responders, and as such be responsible for the safety of the building occupants, passersby, and our own members. It is essential that we understand the dangers and behavior of leaking natural gas and that we take all possible precautions to prevent a disaster from occurring. We need to know what we can do ourselves and when we need help. In those instances that we do need the help of the utility company, we must call for it early.

It is important that we notify the gas utility of all gas leaks, even small ones that we can successfully control. Utility personnel may be responsible for repairing the leak or appliance and will locate or verify the source of the leak while ensuring that no others are present. They can lock the gas service until repairs have been made, thus preventing hazardous, premature restoration of the gas by the occupant. When the appropriate repairs have been made, the utility will certify that the system is safe before restoring gas to the line.

With the advice and assistance of the utility's trained personnel, we can safely, routinely handle the gas emergencies and fires that we encounter, as well as the rare catastrophic incidents. We must, however, never become complacent. Danger is ever present, waiting for us to make a mistake, possibly a fatal one.

STUDY QUESTIONS

1. The natural gas intended for domestic use contains 70 to 90 percent _____ , with _____ making up most of the remainder.

2. Although natural gas is nontoxic, it can cause death by _____ .

3. Typically, the odorant added to natural gas is _____ .

4. Natural gas is about _____ as dense as air.

5. High-pressure gas transmission lines can be pressurized to as much as _____ .

6. Each end user has a gas regulator installed to lower the pressure down to a usable pressure of about _____ .

7. The main danger of having water in a gas line is that it can _____ in the same way that an air slug would.

8. For every increase in temperature of 459°F, the volume of air _____ .

9. To some extent, odorants can be absorbed by iron gas pipe, but only when the pipe is _____ .

10. What device enables you to determine not only that natural gas is present, but also its concentration and whether or not it is within its flammable range?

11. Name the two probable results of excessive gas flow to an appliance.

12. Blocked burner ports, accumulated dirt, and low gas pressure can all result in _____ .

13. Don't extinguish a gas fire until you _____ .

14. Could keying the mike on a portable radio be enough to set off an explosion at a response to an interior gas leak?

15. When responding to an interior gas leak, you should position apparatus defensively against _____ and _____ .

16. True or false: The hazards associated with a propane gas leak are different from those of a natural gas leak.

17. In the case of a slight leak, is a gas indicator likely to give accurate results if you slip the probe underneath the door of a locked apartment? Why?

18. The shutoff found on the supply side of a gas meter is known as a _____ .

19. True or false: Since metal pipe is conductive, gas leaking from a metal pipe poses a static electricity hazard.

20. Is it possible to stop the flow of gas from a small plastic pipe by folding the pipe over on itself?

Chapter Four
Water Leaks

It was nine o'clock on a cold November night in New York City, and we were responding to a call for a water leak in the street. As we rounded the corner and approached the location, we weren't prepared for what we saw. A river of water was gushing across Fifth Avenue at 97th Street and cascading down into the basements of several apartment buildings in its path. As I watched in amazement, a garbage pail floated past. Our lieutenant, realizing the potential hazards that this situation presented, immediately called for help.

The water was coming from Central Park. The only source that could supply such a volume was the reservoir located in the center of the park. A large pipe must have failed. There were five of us on the ladder truck, in addition to the lieutenant. He dispatched two of us into the park to locate the source of the leak and to search for people possibly trapped in their cars. Cutting through a hilly section of the park, 97th Street was creating a makeshift canal for the escaping water. Walking along the higher ground, we came to a depression in the road where the water had formed a small lake. In it, several cars were stranded, the water up to their windows. The drivers stood on the opposite side of the roadway, watching as the water level rose on their cars. Dripping wet, they had all exited their vehicles and scrambled to safety. We notified our officer of their presence and continued to search for the source of the leak. By now, other units were arriving on the scene, and we had plenty of help.

Back on Fifth Avenue, the arriving chief ordered firefighters to check the cellars of all the involved buildings for trapped occupants, as well as any potential hazards caused by the flooding. Others of us, meanwhile, were ordered to check the interiors of the flooded vehicles. Doing so involved wading waist deep into the flood and groping around inside the cars. Luckily, none of us found anyone in distress.

If the water were to continue to rise in the cellars, the electrical service of the various buildings would be affected. The chief had already put out a call to the water department and the electric company for assistance. When these emergency crews arrived, water company personnel were able to locate and shut a valve, stemming the flow of the water. The crews from the electric company identified which building services might need to be disconnected and stood by awaiting orders from the chief.

The source of the flood had been found and shut down, but we were still left with problems. Several basements were flooded. Although the water didn't reach the height of the electric service, the heating units were swamped and couldn't be restarted until the water was removed and the units serviced. There was also the matter of the lake on 97th Street. In addition, gas and oil burner flames had been extinguished by the rising water. We shut down the burners, cut their fuel supply, and notified the gas utility.

You may never be called to a water leak of such magnitude or one with such a potential for loss of life and property, but you will be called to water leaks. In this chapter, I will discuss several of the most common types and describe fire department operations needed to control them.

LEAKS IN BUILDINGS

Leaking Ceiling

We are often called by irate tenants complaining of water pouring through the ceiling. The cause can range from an overflowing sink to a burst pipe to a flooded roof. In any such case, we have to locate the source of the water, stop the flow (if possible), and remove any hazard created by the water.

Flooded Roof

This condition can occur on a flat roof in any type of occupancy. The cause is usually a clogged roof drain. The results can range from an annoying drip to a heavy flow into the occupancy below, depending on the integrity of the roofing material. When responding to these types of calls, there are several things to consider before taking any action.

Is There a Parapet Around the Roof? A parapet that runs all the way around a building can form an enclosure. In the event of a clogged drain, a parapet can create a swimming pool on the roof.

Where Are the Drains Located? Small, flat roofs usually slope front to rear or side to side, whereas larger flat roofs slope toward drains that are typically located at the low points. If there is only a small flood on the roof, the drain will usually be within the flooded area. If the entire roof is flooded, it will be more difficult to locate the drain. Finding the drainpipe will help you locate the drain, but these pipes often run into the building and not down the side. If you cannot find the drain, contact the building maintenance personnel. They should be able to help.

How Much Weight Is Sitting on the Roof? Freshwater weighs approximately 62.4 pounds per cubic foot at 40°F. A flat roof that is twenty feet by forty feet—with a parapet and a clogged drain—will contain 400 cubic feet of water if the water level is six inches deep. The weight of this water will be 24,960 pounds. If the water is a foot deep, the weight will be 49,920 pounds. Each one-inch rise will add 4,160 pounds to this roof. If one drain is clogged on a flat roof, water is frequently found pooled several inches deep in the area around the drain. A pool that is three inches deep and twenty feet square weighs 6,240 pounds.

Has There Been Structural Damage as a Result of an Ongoing Leak? Recently I responded to the collapse of a six-story building. For some time, the residents had been reporting water leaks in the roof. As a result of the cumulative damage from the leaks and a particularly heavy rain, the front wall of the building collapsed into the street, bringing with it the floors of the involved wing.

How Much Weight Is the Roof Designed to Carry? Is the roof in danger of collapse? Has there been a partial collapse? If you suspect a danger of collapse, you should be thinking evacuation and a means to drain the roof. Adding your weight to the roof at this time isn't a good idea, especially if you create an impact load by jumping onto it from the parapet or a ladder. If you decide to operate on such an overloaded roof, allow as few firefighters as possible onto it and avoid impact loads of any kind.

How Do You Clear the Drain? Definitely not with your hands. I recently learned of an incident that almost took the life of a building maintenance worker who had tried to clear a clogged roof drain with his hands. He succeeded in clearing the drain, and when the water started to flow down it, the suction pulled his arm into the drain. Luckily, he was able to keep his head above water and to call for help. The fire department responded. Because his arm was pulled so far into the pipe, the firefighters had to cut and remove the section of pipe that was holding him. They then transported him, pipe and all, to the hospital, where he was cut free. Never attempt to free a clogged drain by hand. You might not be as lucky as that maintenance worker. Use a tool.

What If You Can't Unclog the Drain? What do you do about the water if you can't unclog the drain? One low-tech method is to siphon off the water with a length of rubber hose. A kink anywhere in the hose will stop the siphon, and rubber hose will resist kinking. A length or two of one-inch rubber booster line works well for this purpose. Submerge a length of the rubber hose in the water and allow it to fill, then put a kink near one end and hang that end off the roof. With one end of the hose hanging lower than the roof, a flow of water will start when you unkink the hose. An alternate method is to stretch rubber hose up the side of the building and secure the end in the accumulated water. Then, charge the line using either a pumper or a hydrant. Once the line is charged, shut down the pumper or hydrant, and disconnect the lower end of the hose. The water flowing down and out of the hose will create suction, thus siphoning the water off of the roof. Don't allow any of the roof water to flow back into the pumper, because any gravel or other debris carried with it may damage the pumps.

You can also use a salvage pump to dewater the roof. Depending on its capacity, it may do the job faster than hose. No matter which method you use, you should monitor the flow. No one will thank you if you place the discharge lines carelessly and merely transfer water from the roof to the basement.

Overflowing Sinks, Tubs, and Toilets

Often, when called to an apartment for a water leak, the responding personnel find a frantic tenant trying to set buckets under numerous streams of water flowing from cracks in his water-stained ceiling.

Gain Entry. If the leak is in the ceiling of the top-floor apartment, check the roof for flooding. For all others, check the apartment or room immediately above the scene of the leak. This may be as simple as knocking on the door and asking the occupant to check his sinks for a leak or an overflow. On the other hand, it may be a bit more complicated. What if no one is home upstairs? Will you force entry? How much damage will you do? Can you enter through an open window?

If this apartment is on the top floor, the roof is probably flooded.

If you enter through a window, you must consider the possibility that the occupant may be sleeping, drugged, or in some other way incognizant of your knocking. If he sees someone, even a firefighter climbing through a window, he may try to defend himself. If he has a gun, the results could be deadly. The possibility also exists that the occupant is unconscious as a result of a fall or a heart attack. *Not* entering could result in his death.

Locate the Source. This shouldn't be too difficult. If water is leaking from the ceiling in an apartment, then the source is above you. If you see no obvious source in the upstairs apartment but you hear water flowing, look under the sinks and consider that a pipe may be broken inside a wall. Most likely, the source will be an overflowing sink, tub, or toilet.

Stop the Flow. Simply closing the faucet should stop the water from cascading to the floor. Unclog the drain if the problem is one that is easy to remedy, such as debris in the sink or tub. Don't start snaking the pipe to clear the drain, however. You aren't responsible for repairs. If the toilet tank is overflowing, shaking the handle may solve the problem. If not, remove the tank top and reach in to the float, a metal or plastic globe attached to a thin metal rod. Bend the rod so that the float sits lower in the tank. This should stop the flow.

The problem may also be a stuck flushometer. Again, try shaking the handle. Barring that, look for a hexagonal fitting. If it's loose, tighten it. If that does not work, try tapping it with the back of an ax. If you find the problem to be a leak in the flexible tubing under a sink or toilet, simply tightening the compression fittings may solve the problem. If the tubing, flexible or otherwise, is broken, or if the above tactics don't work for a sink, tub, or toilet, look for a shutoff valve under the sink or the toilet's water tank, or even on the floor below. If the source of the leak is a broken pipe and none of the above techniques work, try to wedge a small wooden plug into the opening. A pencil or even a golf tee may work. Wrapping the plug with cloth or tape before inserting it may help to tighten the fit. The plug can be held in place with duct tape.

If that still doesn't stop the water, you may have to go to the cellar to find the control valve that supplies water to that apartment or line of apartments. Before closing it, consider that you may be leaving many people without water. If you can't locate the appropriate valve, listen for the sound of running water. Locating the source will lead you to the proper pipe. You can also feel the pipes. The one with water flowing through it will be colder than the rest (given that it is a cold-water pipe). If you are unable to shut the water to the apartment or line of apartments, you may have to shut down the house main, stopping the flow of water to the entire building. Remember, it may take time for the water to stop flowing if the leak is on the lower floor of a multistory building. Shutting the supply, however, should quickly slow the leak. Position a radio-equipped firefighter at the leak to inform the firefighter at the valve when the water flow slows down, indicating that the correct valve has been shut.

Kill the Power if Wires or Appliances Are Wet. You can kill the power to an outlet or a fixture, to the apartment, or to the entire building, depending on the extent of the water leak and the involvement of the electrical system.

Flooded Basements

Basements can flood for a variety of reasons. A broken pipe, a rising groundwater table, heavy rain, and overflowing rivers can all flood basements. Many natural disasters far outstrip the power of the fire department, so I will concentrate here on basement floods that come from a source within our control—usually a broken pipe resulting in anywhere from a couple of inches to several feet of water.

Should You Conduct a Dewatering Operation? If the answer to that question is yes, consider the consequences of the operation. How many units will it tie up, and how long will it require their attention? How will the operation affect your ability to provide fire protection to the rest of your response area?

For those times that nature overwhelms your resources, you should have in place a mutual aid plan with other departments and agencies that can supply dewatering equipment and additional staff to help you. Remember that your primary goal is to save lives. Will rising water result in the shutdown of vital systems in a hospital or other health care facility? If the electricity in a building can't be turned off and the rising water creates a fire hazard, then lives may be endangered. Removing occupants and stemming the flood will become part of your primary mission.

Do You Have the Appropriate Equipment to Dewater? A salvage pump requires a minimum staffing commitment, since it can be operated by a single firefighter. An eductor—a dewatering device that uses the venturi principle and atmospheric pressure to siphon water past a jet of water supplied by a pumper—requires that a pumper be left at the scene to supply it with water. This device can safely pump water that contains small particles of debris. Several flooded basements in a small geographic area can be dewatered by one or two firefighters left on the scene with several salvage pumps, whereas using an eductor requires that a pumper be placed out of service for fire duty. You might not want to involve your department with dewatering private property during a storm or other natural emergency. Unless life is in danger, your time might be better spent waiting for and responding to any life-threatening situations that occur as a result of the storm. You can restrict yourself to mitigating hazards by removing electricity, shutting down the gas supply, containing any resultant fuel spill, and evacuating occupants. After the storm, when your response load is less, you can return to assist the homeowner in removing the accumulated water from his basement.

Is There a Life Hazard in a Flooded Basement? Ask yourself, might someone have fallen in? A small child can drown in a few inches of water. Don't forget

A water main break has flooded several basements, requiring a major commitment of salvage pumps and personnel. (Official FDNY photo.)

that your personnel are constantly exposed to life hazards at these incidents, many of them hidden. A flooded basement that seems to contain only several inches of water can become deadly if there are pits or levels below the surface of the water. Some oil burners are installed in pits, and these can fill with water. A basement with seemingly two inches of water can contain an oil burner pit several feet deep—more than enough to drown a child or an unconscious adult. As you move forward through a flooded basement, probe the area in front of you with a pike pole before each step.

If the water level reaches an oil burner, it will cause the burner to malfunction. If it reaches the level of a gas burner, it can extinguish the flame and pilot light. If the water reaches an electrical outlet, it can cause a short that trips a circuit breaker or even starts a fire. If the breaker doesn't trip, the area near the submerged outlet can become electrically charged. If the water level rises sufficiently, fuel oil tanks can begin to float, causing breaks in the piping and indoor oil spills. Once while groping around a flooded basement, I felt a tingling sensation that got stronger as I continued walking. I had the good sense to back out at that point. Notifying my officer that I thought the water was electrically charged, we called everyone out of the basement and had the utility company disconnect the power at the exterior meter.

What's the Source of the Water? This may be determined in a number of ways. You may be able to see water spouting from a broken pipe. If the water is odorous and filled with floating toilet refuse, sewage is backing into the home. If the flooding follows a prolonged period of rain, it may be due to a high water table and leaky basement walls or floor. If the building is attached to other buildings, the source of the water may be next door. In vacant buildings, piping is often removed for "mongo," or salvage. In a commercial building, any of a number of processes may, as a result of damage, equipment failure, or human error, be spilling water into the building. In a private home, a failed hot water heater or broken pipe may be the culprit. Noting the direction of the flow may lead you to its source.

What Can You Do About the Source? There's isn't much you can do about an act of God, such as a natural flood or heavy rain, but you may be able to stem the flow from a burst pipe by plugging it or shutting down a valve. In most cases, turn the valve wheel or lever clockwise to close it. If the leak is from flexible metal tubing, you can crimp the tube. If the source is a broken pipe, you may be able to stop or reduce the flow by tapping a wooden plug into the opening. Simply carve the end to fit the opening and tap it in with the back of an ax. If the broken pipe is made out of a soft metal, it may be possible to hammer it closed.

How Do You Get Rid of the Water? First, as mentioned above, decide whether you want to be involved in a lengthy dewatering operation. If the answer is yes, you can use a salvage pump or an eductor. At fires, when excessive water is found to be collecting on the floor, firefighters sometimes break the toilet bowl at the floor to quickly get the water off the floor. The excess water then flows down into the sewage system. In nonemergency situations, there is no need to damage the bowl, but you may remove it in the conventional manner. Another option is to find and open the cleanout plug of the waste pipe. With either of these methods, it's possible for floating debris to clog the opening. If you can place some sort of screen over the hole, you can keep the water flowing. A metal or plastic milk case works well for this purpose, although you may have to weight a plastic case to keep it from floating away.

General Safety Concerns

Electrical Dangers. Not only may firefighters be in danger of electrocution, the occupant may be also, even after the main water problem has been solved. Water that is still in the ceiling, in walls, and on wires and appliances can cause switches, appliances, and even countertops and floors to become electrically charged. This hazard can occur after you leave if you don't take appropriate precautions. You must ensure that the power to all the affected areas has been shut down and that the occupant understands not to restore it until the water has dried and the system has been checked by a licensed electrician.

When removing power, always consider the safety of the firefighter assigned to the task. It isn't prudent to allow a member to stand waist deep in water when tripping circuit breakers or pulling fuses. In such a case, have a utility worker pull the meter or cut the wire from the exterior to remove the power.

Water Leaks • 101

Water is trapped above this plasterboard ceiling. The firefighter is poking a hole with a halligan to drain it out. He may have to pull down an entire section if the ceiling has become saturated.

Ceiling Tiles and Plaster. Weakened, water-soaked ceiling coverings can fall at any time, possibly causing serious injury. Take preemptive action to avoid such a mishap. Poke a sagging ceiling with a pike pole to relieve the buildup of water above it. For plasterboard, use the handle of the pike pole to make the hole, since it is less likely to cause the ceiling to collapse. Once the excess water has been relieved, it may be necessary to remove a large portion of the ceiling to ensure that it won't fall down at a later time. If possible, move or cover furniture before doing so. In this way, you'll do as little damage to the occupant's property as possible.

Elevator Shafts. Water leaking down an elevator shaft can cause the elevator to malfunction. If water has entered the shaft, check for trapped occupants and place the involved elevators out of service. Firefighters should never use an elevator once water has entered its shaft. Water can affect the car's safety circuits and can result in sudden, unintended movement of the system. At a fire in one elevator-equipped building, a firefighter was straddling the elevator car and the hall floor. Unbeknownst to him, water was spilling down the shaft as a result of the extinguishment operation above. The elevator suddenly dropped down the shaft. Luckily, the firefighter fell into the elevator and wasn't caught between the header of the door frame and the floor. The same sort of sudden movement could occur should water from any source enter an elevator shaft.

102 • Responding to "Routine" Emergencies

This is what you may first encounter at the scene of a water main break. When water is flowing up through the pavement, ask yourself how much damage might already have been caused underneath.

WATER-MAIN LEAKS

Water is delivered to our homes through a grid of underground piping. As this grid ages, it's subject to failure, and it's always prone to damage from careless digging. A water system can even be damaged by firefighters who shut down a hydrant too quickly. I recall once hearing a thud as I shut down one hydrant, after which water started seeping up through the concrete. The flow continued until there was a small geyser of water spouting several inches into the air. I had shut the hydrant too quickly, creating a water hammer that burst a water main.

If left unrepaired, a water-main leak can wash away the soil underneath the concrete and asphalt. The result would be eventual collapse of the sidewalk and street. If a cast iron gas pipe runs through the area, it too could be undermined. If it were to break, it would leak gas. Recently in Manhattan, just such an event occurred. A hundred-year-old water main broke. The dirt in the area was washed away and a gas main broke, the escaping contents of which then ignited.

Verify the Source. When you are called to a water main leak, you must first verify that the leak is from the main. Then, ensure the response of the water department. You must determine the extent of the leak and the area affected, and take action to avoid injury to civilians and property.

Water Leaks • **103**

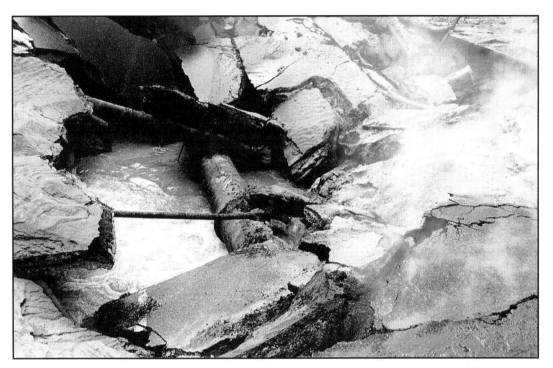

Water seeping up through the pavement at the scene of this water main break has undermined and collapsed a large portion of the roadway, causing other pipes in the area to fail as well. (Official FDNY photo.)

Always notify the gas utility of a water main break. (Official FDNY photo.)

Approach Slowly. At a recent incident in New York City, a firefighter was approaching the scene of a broken twenty-inch water main. Unbeknownst to him, the flowing water had eroded the dirt from underneath the pavement, causing a section of pavement to collapse. The collapsed section, as well as surrounding areas, had filled with water, obscuring the collapse. As the firefighter approached, he fell into the hole and found himself up to his neck in water. He was rescued and was none the worse for wear, though cold, soaked to the skin, and perhaps a bit wiser. The next time he approaches such a leak, he'll certainly test the ground in front of him with a pike pole or other tool before stepping into water of unknown depth.

The possibility always exists that the pavement has been undermined and that the weight of a firefighter or apparatus will collapse it. Naturally, because of the amount of water involved, this danger increases with the size of the broken main. As in the case described above, it's also possible that the ground has already collapsed and that the collapsed section isn't visible due to the water that has filled up the void. Park your apparatus at a safe distance from the break and continue your evaluation cautiously on foot. Just as you wouldn't step off a ladder into a smoke-filled window without first testing for a solid floor, you should not walk blindly into any pool, puddle, or flood if you can't see the bottom.

If the water is flowing aboveground from a large water main break, it may be difficult to walk through the stream. In the incident described at the beginning of this chapter, the firefighters searching vehicles in the flooded roadway had a hard time reaching those cars because the current was so swift.

Another concern is that, under the surface of the water, a manhole cover might have been dislodged or is missing. An unsuspecting firefighter could step into the hole and become trapped in the flooded sewer system. It is important at these incidents, as at any others to which we respond, that we take the time to size up the possible dangers before taking action.

Determine Whether Anyone Is in Danger. If the roadway has collapsed, has a vehicle or pedestrian fallen into the hole? If so, rescue efforts must be initiated. If the water is entering a building, you must consider the aforementioned hazards. In New York City, fire dispatchers automatically contact the gas utility on confirmation of a water main leak. This is by the request of the utility, whose concern is to prevent a major gas leak before it occurs.

Find Out Where the Water Is Coming From. If the water is bubbling up through the street, it's probably a water main break, but if it's flowing from a sewer grating, it may be sewage backing up, and there won't be much that you can do about it. Water flowing from a building out onto the sidewalk normally indicates a problem inside, and you may be able to stem the flow by shutting down the building's water supply. There may be an accessible shutoff located on the sidewalk or in the street. Like a gas curb valve, it will be in a covered metal box embedded in the concrete or asphalt, and it may be marked with a *W* or the word

Water Leaks • 105

Consult with the water department as to which valves you can safely shut down.

water. With the appropriate key tool, firefighters can stop the flow. They must take care not to shut off a distribution main, since this will curtail the supply to a large area. Distribution mains must only be shut down by the water company.

Close Off the Area. Once you have removed all immediate danger from the affected area, you should tape off the scene with fire line tape to keep civilians out, and periodically check nearby buildings for any developing hazards. You must then wait for the water department to arrive and stop the flow from the leaking main. Once the situation has been stabilized and the hazards removed, fire department personnel can take up, leaving the water company in charge of the repair and cleanup.

SUMMARY

Whether the uncontrolled water is from a broken water main or a tub overflowing in an apartment house, the fire department will be called to remedy the situation. In New York City, the fire department is often called, many times in the middle of the night, for a pipe in an apartment that has been leaking for weeks. The tenant, finally fed up with an unresponsive landlord, calls the fire department, in effect, to act as an emergency plumbing service. Although our job is not to fix the leaking pipe, we will stop the flow and, unlike the landlord, we will respond immediately to a call for help. Water leaks can be trivial or they can be truly dangerous. When called to a leak, you must take the appropriate action to prevent injury and property damage, and before leaving the scene, you must make sure that it's safe.

STUDY QUESTIONS

1. Freshwater weighs approximately _____ pounds per cubic foot at 40°F.

2. By weight, how much water will be on a 30- by 45-foot roof if the water level is three inches deep?

3. When siphoning water off a roof, the author recommends that you use a length or two of one-inch rubber booster line since it will _____.

4. One disadvantage of using an eductor is that it requires that a _____ be left at the scene of a dewatering operation.

5. Describe what may happen if rising water meets an electrical outlet.

6. True or false: Dewatering a flooded basement removes the threat of electrocution.

7. True or false: When poking a sagging plasterboard ceiling to relieve the buildup of water above it, you should use the handle of the pike pole.

8. Is it possible to shut a hydrant too quickly? Why or why not?

9. When called to the scene of a reported water main break, what should your first action be?

10. Explain why, if the source of the street flooding isn't obvious, you should search for the location of a broken water main on foot.

Chapter Five
Vehicle Fires

About one-fourth of reported fires are vehicle fires. Insurance fraud, careless repairs, accidents, defects, and arson are all causes of the vehicle fires that keep us busy year-round. Like many other incidents to which we frequently respond, we may eventually come to consider vehicle fires routine, and may not even wear full protective gear when fighting them. This is unfortunate, since each car fire is in actuality a potential haz mat incident. Battery acid, gasoline, rubber, plastic, antifreeze, hydraulic fluid, and all forms of unknown hazardous cargo can be found burning at any "routine" vehicle fire. Bursting batteries, exploding fuel tanks, and carcinogenic fumes are very real dangers, particularly to the unprotected firefighter.

PREVALENT CAR FIRE HAZARDS

Explosion Hazard

There are a number of explosion hazards to consider when determining where and how to attack a car fire.

Bumpers. Today's shock-absorbing bumpers are mounted on compressed gas cylinders designed to absorb the shock of a collision. They perform this job well, but they can also become loaded when the impact compresses the cylinder and it sticks that way. The now-compressed cylinder can suddenly release, becoming a projectile. Two discharging cylinders can blow a bumper completely off a vehicle. If one of the cylinders is blown free and the other remains fixed, the bumper can swing out, hinged on the fixed cylinder. A firefighter in the path of any of these could be seriously injured. At one incident, a loaded hydraulic cylinder was launched from the vehicle and was found embedded in the ground more than forty feet away. To avoid this hazard, approach the vehicle from the sides. Be aware that the cylinder can also explode if it is sufficiently heated.

Batteries. The batteries of today are composed of a plastic shell enclosing lead cells submerged in an electrolyte solution of sulfuric acid and water.

Hydrogen and oxygen are formed as the battery charges. The explosive range of hydrogen is from 4 percent lower to 71 percent upper, and hydrogen can exist in the battery within these limits, mixed with oxygen. If the case of the battery is compromised, an ignitable cloud can emanate from within. If this cloud finds an ignition source, it could flash back to the battery and cause an explosion. The hydrogen gas can develop explosive pressure above 100 psi, translating into a flame front, plastic shrapnel, and a shower of sulfuric acid.

Such an explosion can occur as a result of a battery damaged by impact, fire, or the overhaul efforts of firefighters. The simple act of removing the battery cable can create a spark that will trigger an explosion. To prevent this, remove or cut the negative battery cable, and be sure that it doesn't touch the positive terminal. In some vehicles, particularly diesels, there may be more than one battery, and both must be disconnected. Batteries can also be mounted in the trunk in an attempt to foil would-be thieves.

Tires. When sufficiently heated, the air inside a tire will expand until the tire fails. When it ruptures, it will spew fragments of rubber and steel belts in all directions. If the tire is burning, these fragments will be flaming. Older flat-fixing products once contained petroleum distillates that were highly explosive. Although such products are now required to be nonflammable, many cans of the old type are still in use, and they increase the danger of exploding tires. The force of an exploding tire containing the older, flammable product can be severe enough to raise a corner of the vehicle off the ground.

Flammable Liquids

Gasoline. Although approaching a vehicle from the side may protect you from flying bumpers, it won't guarantee your safety. Approaching a car from the side puts you in the vicinity of exploding tires, and it won't protect you from a failing gas tank. When a fuel tank is subject to direct flame impingement, the gasoline in the tank and fuel lines will begin to vaporize. The resultant pressure can either blow the cap off the filler neck or burst the tank, usually along a seam. A ball of fire fifty feet long, twenty feet wide, and twenty feet high has been reported from exploding automobile gas tanks. A tank will fail at its weakest point, whether at a seam, a connection, or a rust spot. The problem is that you can't know where that point will be at any given incident, nor when or even if the tank will fail. Plastic tanks, used in newer cars, may melt and spill their contents onto the ground, severely intensifying the fire. Fuel lines, too, present a hazard to firefighters. Modern fuel-injected vehicles operate under higher pressure than their old carbureted predecessors. A supply line, which may convey fuel at 30 psi, will retain its pressure when the engine is shut down. A break in such a line could spray gasoline on anyone in the vicinity. In fact, such a cloud of atomized gasoline could be a hazard to anyone close to the car. Fuel-injected vehicles also have a system that returns fuel to the gas tank, but such a line is only pressurized to about 1.5 psi and poses a much smaller threat.

Vehicle Fires • 109

Gasoline is burning underneath this car. Dry chemical or foam are indicated where a hose stream has failed.

Hydraulic Fluid. Both power steering fluid and automatic transmission fluid are flammable when released under pressure in the form of spray. Their flash point is in the range of 410°F to 460°F, and power steering fluid can be pressurized to as much as 200 psi. These flammable fluids are also caustic. They can deteriorate bunker gear and cause chemical burns when absorbed by bare skin. The smoke generated by burning brake fluid is harmful to lung tissue.

Slipping Hazard

One hazard of a vehicle incident that shouldn't be overlooked is that of slippery fluids. Methylglycol mixed with water is used as antifreeze in radiators. Toxic if ingested, it can be troublesome underfoot if it is spilled onto pavement or concrete. Oil, gasoline, and hydraulic fluid all present the same problem, proving hazardous to firefighters as well as to civilians. If the spill occurs on the roadway, passing cars can skid, possibly adding a vehicular collision to your list of woes.

Steam Burns

The temperature of the radiator fluid in a recently operating vehicle is high enough to readily produce steam if the mixture is suddenly released to the atmosphere. This presents a burn hazard to firefighters working near the radiator or hoses. A release of steam could also obscure their vision, thus creating a safety problem.

Smoke

We all know that the smoke at a structure fire is hazardous. Fire departments mandate the use of SCBA at all structure fires, but what about car fires? Do we recognize the hazard of the smoke from burning vehicles? Many of us, even if we do recognize the danger, don't take appropriate precautions. A car fire is a haz mat incident, and the smoke from it contains many toxic elements. Just as structure fires have become more hazardous and the smoke more toxic as a result of the proliferation of synthetic materials, so too have car fires become more dangerous. The available combustibles include plastics, varnish, rubber, lacquers, paint, gasoline, hydraulic fluid, engine oil, and many other natural and synthetic products. These fire-supporting fuels produce thick smoke that contains various hydrocarbon distillates, carbon monoxide, and potassium cyanide, as well as other toxins. Clearly, you must protect yourself. Use SCBA, and stay out of the smoke as much as possible.

ALTERNATIVE FUEL FIRES

With the increased awareness of environmental concerns has come a growth in the use of alternative fuels. Battery-operated cars and those using propane, compressed natural gas, and ethanol are all being used in an attempt to reduce the atmospheric levels of gasoline-bred pollutants. As a firefighter, you must be cognizant of the dangers posed by these fuels if you are to protect yourself from them. Still, at a fire, you may not initially recognize that they are present. A vehicle that uses an alternative fuel may have obvious markings to that effect, or it may not. Even if it does, the markings may be obscured by smoke and flames. Some vehicles even have dual fuel capability, being able to switch from gasoline to propane or gasoline to compressed natural gas.

Propane. The presence of propane and its possible involvement in a vehicle fire pose a real danger to firefighters. A fire impinging on a propane cylinder may lead to a boiling-liquid, expanding-vapor explosion (BLEVE) of the cylinder, with the resultant explosion and fireball. How dangerous would such an occurrence be? The BLEVE of a twenty-pound propane cylinder in one van reportedly propelled sections of the roof as far as seventy-five feet, while blowing the rear doors off their top hinges.

If a propane cylinder is being heated by an impinging fire, and if the fire cannot be extinguished and the tank adequately cooled, the safest tactic may be to pull firefighters and civilians back out of the danger zone. Arriving after a fire has begun, you'll have no way of knowing how long the tank has been exposed to heat or when it may fail. If burning fuel is escaping from the propane tank, you may be able to extinguish the fire by shutting down the fuel supply. One tactic that has been suggested is to approach the fire behind a fog or foam pattern, locate the leak, extinguish it by simultaneously discharging two dry-chem extinguishers onto it, and then reach through the fog stream to close the fuel valve or crimp the fuel line. Obviously, this is a hazardous procedure, and you should only attempt it if there is something to be gained. Remember, this is a car fire. If it is well advanced, there is nothing left to save, neither life nor property. On the other hand, if the explosion of the propane cylinder would put life in jeopardy or if fire is extending to nearby property, attempting to shut it off may be a justifiable risk.

Compressed Natural Gas. Vehicles powered by compressed natural gas (CNG) present problems similar to those posed by propane-powered vehicles. Although a CNG tank can't experience a BLEVE since it contains a gas and not a liquid, it can fail, and such a failure under fire conditions could be deadly to anyone nearby. If such a vehicle is on fire, attempt to shut the fuel valve, but beware of the vent line. It is designed to vent gas in the event of exposure to fire and a resultant buildup of pressure in the tank. If gas vents and is ignited, a firefighter nearby would be engulfed in flames. I am aware of two incidents involving CNG vehicles. One, a pickup truck, was a total burnout. The tank, located in the rear storage area, did not fail. In the second incident, an accident in which a bus demolished a CNG-powered car, the tank also did not fail. I have viewed a video in which the storage tank of a CNG auto was intentionally shot by a bullet from a high-powered rifle. Though pierced, the tank did not ignite. It seems that these types of vehicles don't present any unusual hazards to firefighters, but more experience with them is necessary before such a statement can be made authoritatively.

SEARCHING FOR VICTIMS AT A VEHICLE FIRE

At a structure fire, a primary search is performed during the fire operation, and a secondary search is performed after the fire is under control. These are done in an attempt to rescue live victims and to avoid missing anyone who did not survive the fire. At a car fire, you might think that the victims would be easy to find. Where could they be hidden? A live victim in the passenger compartment will readily be seen unless that area has been crushed by the accident, or unless a pedestrian or motorist is trapped underneath the vehicle. If the passenger compartment, accident or no accident, is fully involved in flames by the time you get there, it's unlikely that anyone in it will have survived such intense heat. Still, you must conduct a search for those who haven't survived.

So what is the problem? Once the fire has been extinguished, won't a fatality be just as easy to see as a live victim? Not necessarily. A badly burned victim may not easily be recognizable as human. In a burned-out car, the seating will have burned down to the springs. The victim may have burned down to the point

It may be extremely difficult to identify a badly burned victim in the charred seating.

that his remains resemble the charred seating. The two might be intermingled, and the human remains might appear to be just so much charred debris. A thorough, well-lighted search must be conducted at every car fire.

An engine company once responded to a car fire off the road in a deserted wooded area. The pumper couldn't get near the car, so several lengths of hose had to be stretched. It was night, and the only lighting available was from the firefighters' hand lights. The car was fully involved by the time personnel were in position with their line. After extinguishment, a search was made, but no victims were found, so the engine company took up and went home. The next day, however, the police department notified the fire department that a tow truck driver sent to remove the vehicle had discovered the charred remains of a victim in the backseat. The firefighters had missed him.

Were the firefighters responsible for his death? Not at all. The car was fully involved when they arrived. No one could have survived the intensity of the flames. It's even possible that the fire had been set to cover up a murder. The firefighters weren't guilty of failing to save the victim, but they were guilty of not finding him. Given the conditions, it's perhaps understandable, but still unprofessional.

A thorough search is necessary at every car fire. Was a murder victim stuffed into the trunk or passenger compartment of a car left in a remote area?

Always search the trunk at a vehicle fire.

Was gasoline then poured onto the victim to incinerate the evidence of the crime? I routinely check the trunks of such vehicles, but doing so is particularly necessary when the burning car is found in a remote area.

Of course, a key would be the best tool to use when opening a vehicle's trunk. If the owner is present, he can give you the key. If not, you will have to force the lock. Driving the point of a halligan tool into the lock cylinder will expose a slot, which you can then manipulate with a flathead screwdriver to pop the lock. If you separate the cylinder from the sheet metal of the trunk by cutting it with an ax or screwdriver, you can then turn the cylinder with a screwdriver, unlocking the trunk. Do not leave a death trap behind when you leave the scene. A child can crawl into the trunk and become trapped in it should the lid close and lock. Bending the striker plate with a tool will prevent this.

CAR FIRE TACTICS

There is no totally safe approach to a car fire, and you must use caution and good sense at all times. Get no closer than necessary, wear full protective gear, and expose the minimum number of personnel required to complete the job expeditiously. Firefighters often rush up to a fully involved car fire with their hoseline and stick the nozzle into the window or engine compartment. In doing

Modern automobiles contain many exotic substances that, when burned, can destroy human tissue. Fighting an auto fire demands prudence and professionalism. This firefighter should be wearing SCBA. (Photo by Matt Daly.)

so, they unnecessarily place themselves in harm's way. Just as we are taught to use the reach of our streams in structure fires, we should do likewise at car fires, at least to darken down and cool the vehicle before approaching it.

Place Apparatus Appropriately. Ideally, the apparatus should be positioned ahead of and at a safe distance from the burning vehicle. What is a safe distance? That depends on the fire. You must consider not only what stage the fire has reached, but also how much worse it can get. One hoseline or fifty feet is a good distance. The idea is to be able to stretch the line and operate the pump safely. You may have to modify this general rule of positioning, depending on where the car is located and other conditions.

If the car is on a slope, facing downhill, positioning ahead of the vehicle would put the apparatus in the path of leaking fluids and even the car itself, were it to roll forward. If the fire occurs on a busy street or highway, you must consider placing your apparatus to block the oncoming traffic, thereby creating a safe working area for firefighters. This may necessitate placing the pumper behind the burning vehicle, but such placement could again make the apparatus an exposure problem if the vehicle happened to be facing uphill. Burning fluids could again flow downhill and expose the apparatus. You must evaluate all factors when determining safe placement.

Choose an Adequate Hoseline. For years, car fires were fought with 1-inch booster line. Today, that is no longer the choice of many departments. Modern vehicles contain more plastics, some have larger gas tanks, and there are other hazards that didn't exist years ago. The increased risk and the increased attention to firefighter safety dictate that you use a larger line, at least $1^{1}/_{2}$ inches, but preferably $1^{3}/_{4}$ inches.

Once you have positioned the apparatus and selected a line, you must then decide how to attack the fire. Other than to save a life, a quick, aggressive attack isn't warranted. Use the reach of the hose stream to darken down the fire while keeping out of harm's way. Dry-chemical extinguishers can be used effectively at car fires. Directing the stream into the engine compartment or the interior will quickly extinguish the fire, but neither will do anything to cool down a dangerously overheated gas tank. Also, a dry-chemical extinguisher won't be able to protect firefighters from excessive heat as a hoseline can.

Passenger Compartment Fire

Knockdown. If a well-developed fire in the passenger compartment has been burning for some time, the windows will have failed, and you'll be able to knock down the bulk of the flames from a distance by using a hose stream. Once you have done this, approach the car from the upwind side and drive the stream into the smoldering upholstery. Then, rotate the stream to hit the roof of the car, as well as the doors and floor. Try to drive the stream into the door cavities and behind the dashboard to extinguish anything that is still smoldering. Banging the side of the door and the roof with a tool will often result in a display of sparks. If so, more water is indicated.

Search and Overhaul. Once the fire has been knocked down, conduct the interior search for victims. At the same time, you should look for possible extension to the trunk and engine compartment. If you suspect an engine compartment fire, you will have to pop the hood for an examination. Likewise, if you suspect fire in the trunk, open it. You may be able to examine the trunk from the passenger compartment. If the backseat has been burned through or removed, you should be able to gain access to the trunk. Quick access for extinguishment can be gained by breaking a taillight and inserting a nozzle into the hole.

Engine Compartment Fire

Knockdown. If the fire is burning in the engine compartment, it's likely that flammable liquids are involved, and a battery explosion is a possibility. It's a good idea to knock down a fire in the engine compartment before opening the hood. This is because the chance of a sudden flare-up of flammable vapors would put the firefighter in danger. You can obtain quick access for a hose stream either

116 • Responding to "Routine" Emergencies

The fire has been darkened down and the nozzle team is moving in to complete extinguishment.

by piercing the grill with a halligan or by breaking a headlight. You could also pry one side of the hood with a halligan or penetrate the wheel well. Cutting the hood with a power saw is also a viable method. Some options available are to cut the hood free from its hinges or its lock, to make a V-shaped porthole at the side, or to render a large X across the center of the hood.

You can also bounce a stream off the ground below the engine. Although not as effective as the above methods, you can do it from a distance, darkening down the flames as you make your approach. Once close enough, you could also apply water by directing a stream into the engine compartment from beneath the car.

Overhaul. Once the fire has been quelled, try to pull the hood release cable in the passenger compartment. This will be the easiest way to open the hood. Often the mechanism will have been damaged by the fire and won't be functional. In that case, reach through the broken grill and pull the release cable directly. If you can't do this, possibly because of damage from a collision, you must pry up the hood with tools, which is no easy task. Once you have opened the hood, prop it open with a tool. A six-foot hook will do the job. Another novel method is to insert the forked end of a halligan over the hinge and bend it about ninety degrees. The deformed hinge in and of itself will keep the hood propped open. Whenever you open a hood, expect the fire to flare up, and position yourself so that you won't be caught by it. Have a handline ready to control the fire.

Beware of Derelict Vehicles. The abandoned, derelict vehicle (or ADV for short) poses several additional problems for firefighters. If left in place for any length of time, it will become a repository for all sorts of trash. In addition, the gas tank may still contain fuel, and all of the other hazards associated with car fires will be present. Some or all of the vehicle's tires may have been removed, and the car may be propped up on milk crates or cement blocks. If any of these supports are disturbed during firefighting operations, the car could fall and injure firefighters. There may also be a life hazard in the car, since children enjoy playing in abandoned vehicles. You will have to conduct a search for a body that could intentionally or accidentally be concealed by rubbish. A fire in an abandoned vehicle is often more dangerous than a standard car fire, and you should approach it with caution.

CAR FIRE IN A GARAGE

A car fire outside of a structure is just that: a car fire. A car burning inside a garage, however, is a structure fire, and all of the rules of structure firefighting apply, as do the hazards of car fires. As in any structure fire, the priority is saving

This vehicle caught fire inside the garage, but the owner pushed it outside and tried to extinguish the flames. (Note the garden hose lying across the roof.) On arrival, firefighters were faced with a fire in the garage and the entrance blocked by the burning car.

life, so if the garage is attached to a dwelling, the first line must be chosen and positioned to accomplish this goal. It may be necessary to stretch the first line into the exposed dwelling.

Stretch an Adequate Line. If you are still using a booster line to extinguish car fires, you may want to reconsider the practice. You definitely shouldn't use one at this or any other structure fire. Use a line that is appropriate for structure firefighting, and stretch it to a location that will protect the occupants of the building as well as the firefighters searching for victims. A 1¾-inch line is a good choice.

As a young firefighter, when we were still using the booster line to extinguish car fires, I once responded to a car fire in a commercial building. We stretched the booster line, but the heat was being contained and amplified by the concrete structure. Unable to extinguish the fire, we dropped the 1-inch red booster line so as to stretch a 2½-inch line instead. The result of changing our tactics midstride was the involvement of several additional cars. When the chief arrived on the scene, he saw the discarded booster line and immediately deduced our actions. The chief soundly chastised the officer for his decision to stretch a booster line into a structure.

A car doesn't have to be inside a garage to threaten a residence. Always consider the exposure. (Photo by Matt Daly.)

Protect the Life Hazard First. You should place the first line between the fire and the life hazard. At a car fire in the garage of a private dwelling, the life hazard is likely to be in the house itself. If there is a door between the garage and the house, stretch the first line to there. Stretch it through the dwelling, not the garage. If that door is open or missing, or if it has failed because of the fire, the residents and firefighters in the building will be at risk and must be protected by the first line.

Decide the Method of Attack. Once the line is in place and charged, you can then decide whether to attack the fire through the interior door or to protect that door with the line and attack the fire through the open garage door with a second line. This exterior, frontal attack is far safer than fighting the fire from inside. The keys to using it are well-trained personnel and a coordinated effort. I know of an instance in which the first-in companies instituted an interior attack on a fire in an attached garage. They correctly stretched into the dwelling and proceeded to attack the fire through the interior door. When mutual-aid companies arrived, they attacked the fire through the garage door. Trying for a quick knockdown, they opted to use their deck pipe. As you can imagine, the interior team was driven back by the smoke, heat, flames, and steam. The result of the uncoordinated attack was substantial damage to the structure as fire extended at multiple points, forcing the evacuation of the interior forces. Clearly, a lack of communication was the crux of the problem. Still, a properly coordinated frontal attack places the firefighters outside in relatively clean air, away from the many dangers inherent to car fires.

If you choose to attack the fire through the exterior garage door, you must quickly check for extension from the garage into the dwelling. An opened or failed door, ductwork, damaged or missing plasterboard, poor workmanship, pipe recesses, or any other opening in the wall can provide a path for the fire. The check for extension from a garage must include the floor or attic above. The interior line must protect the closed interior door. Have enough slack in the line to cover all parts of the building, and be able to move quickly to cover any point of extension. If the fire is extending at multiple points, additional lines will have to be stretched, possibly one to each floor or level of the home.

Garage Fire Hazards

The Car. One obvious hazard is the vehicle itself. We know that outdoor car fires are hazardous, given their flammable liquids, plastics, battery acid, and hydraulic bumpers. All of these become more dangerous when they are confined in a garage. The sudden involvement of a fuel tank in this confined area can injure firefighters, cause structural damage to the garage, and promote rapid extension to the residence itself.

Overhead Garage Doors. The cable and springs that raise an overhead garage door can suddenly fail when they are exposed to high heat. The result can be that the door suddenly comes crashing down into the closed position. If any firefighters have entered via this route, they will be trapped until the door can be raised again. If the line has been stretched into the garage, the falling door can cause the line to burst, leaving those members trapped and without water. If the line hasn't yet been charged, the door will choke off water from reaching the nozzle. (Firefighters shouldn't enter the garage before the line has been charged.) In any case, the members inside are in trouble. Anytime you pass through a roll-down garage door, plan to keep it opened. You can accomplish this by wedging the tip of a six-foot hook into the track or by clamping a vise grip onto the track under the raised door. Be certain that the hook is firmly set. If the ground is greasy, the hook can slip or it can be knocked out of place. I have seen a commercial metal garage door come crashing down onto a firefighter when he dislodged the ladder that was supporting it. The door should be propped up before any personnel are committed to working within the garage. If possible, assign someone to monitor the door and ensure that it doesn't come down. Another dangerous possibility is that the entire door might fall down from its tracks as the tracks are warped by intense heat. If the door drops onto personnel below, it can injure or trap them.

Stored Goods. Take a look inside your own garage. How would you like to fight a fire in it? You, as a firefighter, would never store gasoline in your garage, right? You wouldn't have propane tanks there, right? You wouldn't keep pesticides, lawn chemicals, paint, and lumber there, would you? If you, as a firefighter, have some of these hazards present, imagine what you will find in the garages of ordinary citizens. Rest assured that everyone else's garage is at least as bad as yours. The reality is that *any* fire in a garage is usually a haz mat incident. At haz mat incidents, firefighters are told to go slowly and not to take action until they assess the hazard. In the case of a fire in an attached garage, you don't have that option. You can't know what is burning until you extinguish the fire, and the victims in the house may not survive if you adopt the go-slow standard. So what to do?

Garage Fire Tactics

Beware of Potential Hazards. For starters, you must take all precautions possible to prevent injury or contamination. Full firefighting gear is a must, as is SCBA. When possible, it's safer for you to attack the fire from the exterior, using the reach of the hose stream. Once the fire has been darkened down, take the time to identify the hazards. Is there a leaking container? Can you plug it, contain it, or remove it? After you have assessed the hazards, you can enter the garage for overhaul and final extinguishment. If the garage is attached to an occupied dwelling, attacking from the exterior may not be an option. You should be able, however, to reach the entire garage with your stream right from a position at the interior door. Striking at the flames and alternately sweeping the floor and ceiling should effect a quick knockdown. Attacking from this position, however, may place you in the path of heat and smoke being drawn into the residential part of the building. This is especially true if accessing the garage requires you to go down a level. In such a case, you will be facing conditions akin to those of a cellar fire. Moving down the stairs quickly will get you out of the hot gases and into a more tolerable atmosphere. If you cannot move down the stairs because the conditions are too severe, protect the interior door and have another line attack the fire from the exterior door.

If you encounter stubborn flames, there may be some exotic fuel stored in the garage, and additional water or some other extinguishing agent may be required. White flames that intensify when water is applied could indicate a fire involving magnesium, a metal used in some auto parts. In such a case, large amounts of water will be required.

If a hoseline won't put out the fire, consider using a foam line or a dry chemical extinguisher. If burning flammable liquid is flowing out of the garage, cover it with foam and dike it, if possible. It may be necessary to check any buildings, drains, and sewers in the path of the leaking fluid for flammable vapors.

When Possible, Vent the Garage. If the attack is being made through the interior door common to the house, opening the main garage door will provide ample ventilation. If the attack is being made from the exterior, vent the garage through a side or rear window or door, if present. As in all structure firefighting, the idea is to vent the fire ahead of the advance. If you cannot effect proper ventilation ahead of a hoseline advancing from the exterior, the open main garage door itself may suffice, allowing smoke and heat to vent over and around the nozzle team. Smoke, heat, and flame may be driven into the dwelling or out of vent openings. In addition to the line stretched through the interior, it may be advisable to stretch a precautionary line at the vent location if fire emanating from a vented door or window would expose the siding, upper-floor windows, or other buildings.

It isn't practical to vent a garage roof during the early stages of a fire for several reasons. First, the garage may be beneath the home, with a bedroom over it. Venting the roof wouldn't vent the garage. Initially, staffing will be better put to use stretching the attack line and searching the dwelling for victims and extension. If the garage is detached, its roof rafters and deck may not be substantial. As a result, they may fail early if they are involved in heavy fire, endangering firefighters assigned to ventilate the roof.

OPERATIONS ON HIGHWAYS

Responding to car fires on highways and other busy roads can be hazardous to firefighters as well as to civilian drivers. High speeds, rubbernecking, heavy traffic, blinding smoke, and driver frustration can all result in additional mishaps at the incident scene. When you respond to a car fire on a highway or a busy street, you must take action to protect yourself, as well as the unsuspecting drivers that you will encounter. This type of response should not be a single-unit operation. Safety dictates the response of at least two units.

Warn Traffic. Assign one or more firefighters to divert or stop traffic. They should be dressed in reflective firefighting gear and have hand lights at night. Place flares well ahead of the area where firefighters will be working. The distance that flares should be from the incident site depends on the speed of the traffic. You must give oncoming traffic enough time to perceive the danger and react to it by stopping or slowing down. According to the American Automobile Association, the stopping time of a car depends on the perception time of the driver, plus his reaction time, plus the braking distance. It will take time for a motorist to perceive any cones or flares that you set up, and it will take time for him to react appropriately.

APPROXIMATE STOPPING DISTANCES FOR PASSENGER VEHICLES

Speed	Stopping Distance	Reaction Delay	Total Stopping Distance
20 mph	18–22 feet	22 feet	40–44 feet
40 mph	72–88 feet	44 feet	116–132 feet
60 mph	162–202 feet	66 feet	228–268 feet

In addition to these criteria of braking distance, there will always be the variable of perception time. Many factors affect perception time, including visibility, the condition and age of the driver, and the nature of the road. Perception time varies with each driver and can even vary for the same driver at different times.

For sixty-mile-an-hour traffic, place the leading flare at least 268 feet ahead of the incident scene, and park the apparatus defensively, if possible. In the configuration shown below, the pump operator is shielded from oncoming traffic.

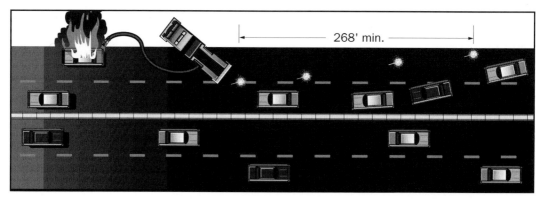

You should space out flares beyond the estimated stopping range of oncoming traffic and in such a way as to direct the traffic away from fire operations. For example, if the burning vehicle is located in the right lane of a multilane highway, the flares should start against the right shoulder and gradually slant left, toward the farthest lane occupied by emergency personnel. When possible, place the apparatus ahead of the burning vehicle, behind the last flare in the series. This way, you will have a protective barrier between you and the oncoming traffic. When placing flares, consider the path that leaking gasoline might take, and do not allow the flares to become sources of ignition.

Request Assistance. Request police assistance early. The police can control the flow of traffic for you. Consider closing off the road to all traffic. This won't sit well with the police, since they are responsible for keeping the traffic moving. In one instance that I am aware of, a state trooper arrested a fire chief because the chief wouldn't open up one lane of a highway where firefighters were operating at the scene of an accident. The chief was concerned with safety, the trooper with moving traffic. Who has the authority to make that decision in your locality? Find out before your next response. Still, if conditions are such that traffic is a hazard to either firefighters or civilians, do not hesitate to shut down the road.

Stay Off the Roadway. When possible, conduct all firefighting operations from off the road. Even if the vehicle is located on the shoulder, keep your firefighters off of the right of way. You should still block the traffic lane closest to your operation. This will give you added protection from traffic. At the same time, it will distance traffic from the burning vehicle—a safety measure in case the vehicle explodes.

Place Apparatus Defensively. When operating with only one apparatus, position it between your firefighters and oncoming traffic, if possible. This will create an effective barrier. It's better that an oncoming car hit the apparatus than one of your members. All of the firefighters should operate behind the barrier. When more than one unit is operating, place the operating pumper past the car fire and the second apparatus between the oncoming traffic and the burning vehicle. If the fire is in the middle lane of a multilane highway, it may be advisable to block off more than one lane of traffic; otherwise, cars will slip past you on both sides.

As soon as practical, chock the wheels of the burning vehicle. This is important whether the ground is level or sloped. Of course, if the vehicle's brakes fail when it is on a hill, it can roll. Yet, it is possible for an engine to start as a result of a short caused by the fire or firefighting efforts. If the car has a standard transmission and is in gear, the engine could cause the car to lurch forward, injuring firefighters in its path.

Adjust the placement of apparatus for hills and blind curves. Positioning a pumper just past the crest of a hill can make it invisible to oncoming traffic, as can placing it around a blind curve. If you must operate in such locations, position another apparatus where it can easily be seen by oncoming motorists. Also, place flares in advance of the curve or the top of the hill.

Perform a Proper Cleanup. If the car fire resulted from a collision, it's likely that automotive fluid of one type or another has spilled onto the ground. Even if there was no accident, it's possible for fluids to have been released by the fire or firefighting efforts. In freezing weather, the road will be slick with ice formed from the water discharged from the hoseline. If you simply put the hose back on the apparatus and depart, you may be leaving a potentially hazardous situation behind. It's likely that the spill contains gasoline, oil, antifreeze, transmission fluid, brake fluid, or some combination of them all. Such a spill will be as slippery as ice, and as cars drive over it, they will spread it over a wide area. It's just a matter of time before an unsuspecting motorist tries to stop or turn but instead slides out of control, possibly causing an accident.

A little bit of absorbent or sand can prevent the above scenario. If the spill is small, it can easily be absorbed by a commercial absorbent or a few shovelfuls of dirt or sand. For a larger spill or an ice condition, you will need help. I often make use of a sand spreader from the sanitation department if fluids have spilled or if suppression efforts have created ice on the roadway. Such a truck can shoot sand over the entire area of the spill, even underneath the involved vehicles. Firefighters with shovels can cover any areas that are missed. Remove any large or sharp debris from the roadway, since it might cause an accident. If the vehicle was involved in criminal activity, the police may declare it a crime scene and won't want you to move anything. If the vehicle is going to be towed, consider leaving a pumper on the scene as a precaution until the car has been hooked up and is on its way.

SUMMARY

Vehicles today are more hazardous to firefighters than they were in the past. They contain more combustibles, and when they burn, they emit more toxins. Firefighters must take appropriate precautions whenever confronted with fire in these rolling haz mat carriers. Masks, full bunker gear, and hoseline of the right size should be mandatory. In most cases, a slow, deliberate approach is best, using the reach of the hose stream to keep you out of danger.

Determine whether there is a life hazard in or around the car, then plan your attack accordingly. If there is no life hazard, don't place your own life at risk. If there's an exposure hazard, simultaneously protect the exposure and attack the fire. When possible, place the line between the vehicle and the exposure, and alternate between them. Stretch a second line if needed, and as soon as possible, check the interior of the exposure for extension.

Most of all, fight the tendency to regard these types of incidents as routine, and try to see them instead as the biological, ecological, and pyrolytic hazards that they are.

STUDY QUESTIONS

1. A firefighter should treat every vehicle fire as a potential _____.
2. To avoid the hazard of shock-absorbing bumpers, how should you approach a burning vehicle?
3. What two gases are formed within a battery as it charges?
4. The explosive range of hydrogen is from _____ to _____.
5. In some vehicles, particularly those with _____ engines, there may be more than one battery.
6. What is the principle hazard associated with the fuel lines of modern fuel-injected vehicles?
7. Are power steering fluid and automatic transmission fluid flammable?
8. A fire impinging on a propane cylinder may lead to what type of explosion?
9. Can the fuel tank of a vehicle powered by compressed natural gas experience the same type of explosion as a propane tank?
10. If, after a fire in a vehicle has been put out, a thorough search of the interior and trunk of the vehicle reveals no victims, is the search complete?
11. The author recommends thoroughly searching the trunk at every incident, but particularly under what condition?
12. Normally, if the burning vehicle is on a slope, you should place the apparatus _____ of it.
13. Name two ways to introduce water into the trunk besides opening or cutting the lid.
14. Treat a car burning inside a garage as a _____.
15. How can a vise grip be used to hold up a garage door?
16. When setting out flares to warn traffic of an incident site, how should they be placed?

Chapter Six
Kitchen Fires

"It smells like food." That is a radio report that I hear several times a tour. Ordinarily it means that someone has left a pot on the stove or a roast unattended in the oven. Such a call often requires fire personnel to do little more than turn off the stove and place a cover on a flaming pot, or place a smoldering pot in the sink. Besides the familiar smell of burning food, white smoke drifting out the kitchen window typifies these fires.

What is being cooked can affect the type of incident you encounter. *Fire Findings Journal* conducted tests on corn oil and beans to see how they react to prolonged exposure to heat. Corn oil in a frying pan was left on a gas stove over a high flame. The oil started to smoke in ten minutes and ignited in twenty-two. Two minutes after that, the flames were tall enough to touch an eight-foot ceiling. (See "Kitchen Fires: It Pays to Know What's Cooking," *Fire Findings Journal,* Summer 1997, p. 13.) In another test, beans were heated for seventy minutes. Despite reaching 303°F, they didn't ignite. Although the beans did give off a good amount of smoke, the burning oil produced more.

FOOD ON THE STOVE, NO EXTENSION

We record every food-on-the-stove incident as a structure fire, even if the fire never spreads from the pot. If you were to respond to such a call and beans were the culprit, it's likely that there would be no extension from the pot. However, you would likely have to contend with heavy smoke. You could expect the residence to be filled with nasty-smelling smoke that would make you hack and cough. It would obscure your vision, and until you vented the interior, you would be operating in near-zero visibility. You would have to gain entry, search for possible victims, find the stove and shut it off, ventilate the area, and search for possible extension.

Tactics for Food-on-the-Stove Fires

Gain Entry. If the person placing the call is the occupant, then you should have no trouble gaining entry. He should be able to let you in with a key. If the occupant isn't the calling party, then he may open the door a crack and swear that there is no fire, even as smoke pours out from behind him. It isn't uncommon for a careless cook to be too embarrassed to admit his mistake. Recent immigrants may fear punishment, arrest, or expensive fees if the fire department responds to their home. As a result, they may deny that the source of the smoke is from within their residence.

Two common reasons for food burning on the stove are that an occupant has gone out and left food unattended or has fallen asleep while cooking. If he is home, banging on the door might awaken him. If he isn't home or doesn't respond, you must gain entry.

There are two schools of thought on forcing entry for food-on-the-stove calls. One is that you should use the least damaging option, even if it delays entry. The reason given is that burning food will likely do little real damage and that you shouldn't add to the damage by destroying the door. According to this school of thought, a through-the-lock method, waiting for a key from a neighbor, or climbing in through an open window are all preferable to damaging the door.

This makes sense if there is no fire, only smoke, and when no one is in danger in the residence. The obvious dilemma is that you can't be sure that the

A firefighter enters the kitchen window from a ladder. No smoke is pushing out of the apartment, but there is an odor of burned food evident in the hallway and at the window. The officer in command has opted to do as little damage as necessary for what is apparently a minor food-on-the-stove emergency.

smoke is just from beans and that oil hasn't ignited, or that the occupant hasn't been overcome by the smoke or suffered a heart attack. If you opt for the least damaging method of entry, you may be giving a small fire time enough to ignite surrounding combustibles. Once the oil ignites, it's likely that the flames will involve some nearby combustible, and then you'll have a more serious fire with which to contend. This will necessitate water and overhaul, leaving more damage than would have been caused by forcing the door.

Another consideration when deciding how to gain entry into an occupancy is the safety of the firefighter. Should a firefighter climb up a ladder and into a window, possibly risking a fall? Should he climb down an ancient fire escape to save the expense of repairing a door? What if a sleeping or drugged occupant, startled into consciousness, mistakenly shoots the firefighter? What if the occupant is deaf and encounters the firefighter in a smoky hallway? He may act to defend himself, possibly injuring the firefighter. What if the window is booby-trapped to keep out burglars? In high-crime areas of New York City, some windowsills and security gates are electrified to thwart burglars. The means by which firefighters can be injured are almost endless.

The decision as to how and when to gain entry isn't a cut-and-dried matter. One senior firefighter whom I knew always immediately forced entry for any food-on-the-stove call to which he responded. He explained that once, early in his career, he was called to such an incident and that it took a long time to gain entry. He and his crew members first tried the fire escape, but those windows were locked. Then they tried using a portable ladder, but all of the other windows of the apartment were locked, too, and they didn't want to cause unnecessary damage. Finally, they pulled the lock from the door and gained entry, only to find the occupant lying on the floor, unconscious, as a result of a heart attack. The victim subsequently died, and the firefighter never forgot the incident, because he believed that the victim might have been saved had they taken quicker, more decisive action.

A scenario that I have frequently encountered is being called to a home or apartment by an occupant because he is locked out. When we arrive, we find that the occupant wants us to gain entry for him. In many cases, the call specifies that the occupant has been locked out and has left food cooking on the stove. Sometimes the caller reports that a child is locked inside with burning food. Often, after gaining entry, we find no pot, no burning food, and no small child. They were fabricated to make the call seem more serious and to ensure our response and assistance.

In fact, the person standing there may not be the occupant at all, but a burglar wanting us to do some of his work for him. In such a case, if you decide to enter a locked apartment, you should verify the identity of the caller. Check his ID or ask his neighbors. If you cannot verify his identity, call the police department and ask them to do so. Once you put in a call to the police, a bogus occupant may well disappear.

Which, then, is the correct course of action when responding to a report of food burning on the stove? Should you damage the door or find another, less damaging way in? The answer, I'm afraid, is that the situation entails a judgment call that will have to be made on a case-by-case basis. Is the smoke heavy? Is heat coming from the window? A light haze might indicate a less serious incident. Heat from a window would indicate fire extending to combustible furnishings or structural elements. Do you know whether anyone is home? Is there a car in the driveway or toys on the porch? If you think someone is home and no one responds to your banging, the likelihood that someone is in distress increases. On the other hand, if you always force entry, you needn't ask these questions, but you may have to explain a damaged door to an irate homeowner.

Search for Victims as Well as the Source. As in any incident, saving life is the primary concern. On gaining entry, you must simultaneously search for the source of the smoke and for any unconscious victims. If entry has been gained through a window, a single firefighter may perform the initial search. In the process, he may let other personnel in through the door, provided that door does not require a key. If entry has been gained through the door, several firefighters may join in the initial search. Light smoke means an easy, visual search. Heavy smoke will entail conducting the search as in any structure fire: by touch. After the smoke lifts, conduct a thorough visual search of the occupancy.

If the incident is in a multifamily building and the smoke is heavy, the public hall will fill up with smoke when you open the apartment door. Remove any occupants from the stairs and public hall before you open the door. You can usually vent the hallway and stairs by opening the bulkhead, scuttle, or stairway windows. A smoke ejector may be required if the smoke won't lift because of the weather conditions or if the smoke is trapped in a long, unventable hallway, as can be found in some high-rises. Not venting this smoke can result in additional calls from the occupants after you leave the scene. Actually, smoke in the hallway of any multifamily building can result in multiple calls to the fire department. If you are notified by your dispatcher of additional calls from the same building, you'll have to check out each one to ensure that there is no danger. Remember, it's possible that the additional calls might not have been prompted by the incident at which you're operating. An entirely different smoke condition might exist. You must check out each call.

Consider the Life Safety Issues. Any victims will have to be evaluated and, if necessary, treated. Is the victim unconscious? Is his condition the result of the smoke, did he have a heart attack, or is he under the influence of alcohol or drugs? You must assume the worst. Summon an ambulance to ensure that he receives the appropriate medical assistance.

On the subject of life safety, consider the risks to the firefighter. Many risks are present at such incidents. The obvious danger is the choking smoke that accompanies many food-on-the-stove incidents. Firefighters who have responded to such calls are familiar with the nasty smoke they often find. It saturates their clothes and hair, and the smell stays with them for the rest of the day. Smoke from burning food probably contains sulfur, nitrogen dioxide, and carbon monoxide. Sulfur is well known to be an odorous substance. Nitrogen dioxide can cause nausea, stomach pains, and vomiting. Since it enters the bloodstream, it can also create an oxygen deficiency in the blood and alter blood pressure, thereby causing headaches. These reactions can be delayed, appearing well after the exposure. Carbon monoxide is deadly in sufficiently high concentrations, of course, but it's also suspected to be a cause of heart disease when there is prolonged exposure to apparently harmless amounts. In simple terms, food smoke can be irritating and difficult to tolerate. As far as I'm concerned, anything that feels so bad can't be good for you. In addition, the metal pot, plastic handle, and nearby combustibles, if ignited, can be contributing their smoke to the mix. Toxic gases can be generated before active flaming starts. Even the heating of a plastic handle can generate toxins. Wearing a mask might make it more dangerous to climb a ladder and enter a window, but you must still wear one.

Another danger to the firefighter is the hot pan and its contents. The pan itself, in less than ten minutes, can reach temperatures above 900°F. Grabbing it with an ungloved hand would result in serious injury. If the contents were hot oil or another liquid, a spill might burn a firefighter even through his bunker gear. Applying water to burning oil might cause it to splatter.

If a gas flame is inadvertently extinguished without shutting down the gas, an explosion is possible as the incident escalates to include a gas leak. You should also consider the presence of strong cleaning chemicals in the vicinity of the stove. Kitchens commonly contain bleach, ammonia, pesticides, oven cleaner, drain cleaner, and a host of other chemicals that can become problems if they spill or become involved.

Ventilate. The ventilation requirements will be different at each incident, depending on what is burning, how much smoke has been generated, how large an area is involved, the type of building, and the weather conditions. In most cases, you needn't inflict damage to effect adequate ventilation. Opening double-sash windows two-thirds at the top and one-third at the bottom will amply serve to clear an apartment. Breaking glass usually isn't warranted for food-on-the-stove incidents that don't escalate into structure fires.

If the windows are of the newer, energy-efficient type, they can often be partially or completely removed by detaching them from their tracks. This allows you to open nearly the entire window area for ventilation. If smoke has spread into the public halls, it may be necessary to open the roof door or scuttle, if one is present, or to open the hall or stair windows. In a high-rise, it may not be possible to vent the public halls because of a lack of windows. Use a ventilation fan in combination with an opened apartment door and window to vent these areas. The smoke generated from a food-on-the-stove incident will cool down quickly and, as a result, may not vent easily. Opening windows on more than one side of the structure to create cross-ventilation will usually suffice, but if it doesn't, a fan may again be necessary.

The firefighter who initially enters the residence by way of a window to search for victims and the source of the smoke should vent as he searches. If he encounters heat in the apartment, he must be careful not to pull fire toward him as he vents. Heat is a sign that this is more than a food-on-the-stove emergency and that the rules have changed. If the firefighter encounters heat, he is now searching a burning occupancy without the benefit of a line and must exercise due caution. Part of that caution includes calling for a hoseline. In this situation, breaking glass is warranted if it will assist him in his search for life. If heat is evident immediately on entry, he should call for a line at once and the door should be forced. Stretch the line to the door and charge it.

Locate the Stove. In private homes, the kitchen is often located in the rear of the house on the first floor. Often it will have an exterior door of its own. In apartment buildings, the kitchen can be located anywhere. In New York's so-called railroad flats, apartments that ran front to rear for the entire length of the building, the kitchen was found in the rear. In newer apartment buildings, the location of the kitchen is usually the same from floor to floor—i.e., the kitchens will be aligned vertically. You should know where the kitchens are commonly located in the buildings in your district, and this is the direction you should go once you gain entry.

It's possible for a home to have more than one stove. In some areas, it's common to have a second stove in the basement to help prepare large meals at holiday time. If you encounter food smoke but find no food on the kitchen stove, you'll have to expand your search. Food might be burning on a second stove, in a microwave, or on a hot plate anywhere in the dwelling. There may be an illegal apartment in the building, and the source of the smoke might be there. Recently we found a second stove in the attached garage of a two-family home. It was being used to cook a large amount of fish that had been caught by the occupant. The occupant had left the stove on and departed the home. When smoke alarms sounded in the top-floor apartment, we responded and forced entry into the first-floor apartment. Finding a medium smoke condition, we also forced a door that opened into the garage. We then searched the entire house and finally returned to the garage, where we found the ancillary stove. We had initially walked right past it in the heavy smoke. No one had thought to look in the garage for a stove, but the lesson is clear: You must check everywhere.

Mitigate the Problem. The typical kitchen incident is solved by putting the smoking pot in the sink and either filling it with water or covering it with a lid to smother the flames. At a recent incident, a firefighter placed a burning pan of oil into the sink and turned on the faucet to extinguish it. The water turned to steam, and the hot, vaporized oil rose with the steam cloud, causing a three-foot flame to lick up from the pan. It happened again when the firefighter tried to extinguish it with water a second time. Final extinguishment was effected when the stream from a water extinguisher was applied from a safe distance. This firefighter learned that it might be better to cover a burning pot to extinguish the flame and then to place it into the sink and allow it to cool naturally. If the pan continues to generate smoke, consider removing it from the building. Be careful not to spill hot or flaming oil onto yourself, furniture, or other combustibles as you carry it outside. One of your first actions should be to shut off the stove. Then extinguish the fire, if present, by covering the pot to smother the flame.

Search for Extension. If the pot is just smoking, it isn't likely that fire will have extended to furnishings or the structure. Still, you'll have to make an examination to be sure. Were there flames prior to your arrival, and did they scorch the underside of the cabinet above the stove? It may be necessary to scrape the char or blistered paint from the cabinet to see whether anything is smoldering beneath it. Flames from a pan with several inches of burning oil can burn themselves out in ten minutes. All you'll encounter by the time you arrive is the residual smoke. Did the flames ignite anything before the fuel was consumed, perhaps a pot handle, wooden spoon, roll of paper towels, or possibly the contents of the cabinet? Is something else burning on the stove or possibly in the oven? Check to be sure.

A fire that started in a toaster oven scorched the kitchen cabinets and required extensive overhaul.

A food-on-the-stove call can turn out to be anything from a minor annoyance to a major conflagration, and you won't know which until you perform a proper reconnaissance of the site. The odor of food can be deceiving, since it may be coming from another, unrelated incident. You must always check the reported occupancy and also check with the location of the calling party. If you don't, you might be sidetracked by an unreported, food-related nonincident on the first floor and miss the mattress fire on the fourth floor. You know that if you can see smoke, you have something, and what appears to be only a burnt pot can be more involved. If the burning food was accompanied by active flaming, you must make a thorough examination for extension. If you can't extinguish the flames by simply covering the pot, you may have to use an extinguisher or stretch a line and apply water. If the fire has spread from the stove, you may have to perform overhaul. If it obviously isn't a minor incident, it's advisable to stretch a precautionary line.

FOOD ON THE STOVE, WITH EXTENSION

The flames from the burning oil rose five feet in the air. They lapped around the underside of the wood-and-plastic cabinet above them and into the open cabinet door. The cardboard and plastic contents of the cabinet ignited. Plastic

A seemingly minor stove fire can result in substantial flame spread.

cups melted. The window curtains, blown over the stove by the wind entering the open window, ignited and quickly burned. Heat and flame entered a hole in the soffit above the cabinets, igniting the framing wood. When the oil first burst into flame, the careless cook had run out of the apartment and called the fire department from a pay phone down the street. On arrival, the firefighters were faced with heavy smoke boiling out all of the open windows in the apartment. The cook told the firefighters that it was only a pot burning on the stove and begged them not to damage his apartment. This may have started as a food-on-the-stove incident, but it's a structure fire now.

Anything flammable left on the stove can be ignited by a careless cook. A pot holder, towel, or stack of napkins can be ignited by a stovetop burner. Food left too long over a heat source can burst into flames. Built-up grease in an oven can ignite. Plastic handles on pots can melt and in some cases ignite. Spilled oil or grease can drip onto the floor, into a nearby garbage pail, or into a bag of groceries. A bag of garbage inadvertently left on a stove and ignited by a pilot light was responsible for a fire that claimed the lives of three New York City firefighters. All of these sources can generate smoke. Once the fire spreads beyond the stove, the building contents and possibly the structure are involved.

Sometimes the cook will try to extinguish a stove fire himself, often with disastrous results. Spilled, flaming oil extends the fire and burns the cook; or the cook extinguishes the gas flame and the leaking gas ignites explosively; or the cook gives up and runs out of the occupancy, leaving the front door open, allowing a free passage of air to fan the flames. Given any of these events, arriving personnel will be greeted with a structure fire, and they must go into action for such.

Tactics for a Stove Fire That Has Extended

Gain Entry. Confronted with a structure fire, there is no longer room for discussion as to what method to use to gain entry. Enter the premises, and do so quickly. Any damage to the door is inconsequential, since delay could result in a loss of life or of the structure.

Search for Life and Fire. You must search for both life and fire simultaneously. Quickly remove trapped victims to safety and give them medical attention. Find and extinguish the fire. Moving rapidly to the kitchen may confine the fire to that area. If the structure is a multifamily or multistory building, you'll have to search the adjoining units and the floors above. Check for fire extension on the floor above and the adjoining apartment at the kitchen pipe chase. If it is a top-floor fire, you may have to check the cockloft.

Extinguish the Fire. A $2^1/_2$-gallon water extinguisher may be able to contain or even extinguish the fire while you wait for a line to be stretched. In any case, the line should be stretched and charged as a precaution. Just because this fire started as a food-on-the-stove call, it's no less dangerous than any other fire and

136 • Responding to "Routine" Emergencies

mustn't be taken lightly. You can put out a fire in an oven by adding a small amount of water to the oven and closing the door. A glass of water or a short burst from an extinguisher will probably be enough to do the trick. The steam that's generated after you close the door will smother the fire. Of course, you must first shut off the gas; otherwise, you'll wind up with a potentially explosive situation.

Ventilate. If the extension is truly minor, you may not need to break any windows. Opening them may suffice. If there is substantial smoke and heat, vent as you would at any structure fire. The reduced heat and better visibility will make your job easier and safer. The smoke will contain more heat than the smoke from the food-only fire, and it should easily rise and vent from the windows without help from fans.

Perform Overhaul. You may have to do substantial damage to the structure if you think fire has extended to it. For a fire in a built-in wall oven, you may have to check around all sides of the oven for extension. This will necessitate removing the nearby wall covering and cabinets. If the wall or cabinets are charred or if the paint is blistered, scrape them with the adze end of your halligan or ax. If the char goes deep, you may have to open the wall or pry apart the cabinet to check for extension and to open up a space for applying water. Look

A child left this stove unattended and the resulting fire caused extensive damage.

Kitchen Fires • **137**

Removing the soffit over these charred cabinets revealed ductwork and other likely avenues of extension.

into the cabinet. Is anything ignited or smoldering? The boxed-out area over the cabinets, the soffit, can allow for fire travel to areas remote from the stove. If you think fire may have extended to it, open it and examine the space. Depending on conditions, you may have to open the ceiling to check for fire. Check first where the ceiling is pierced, since this is where it will be easiest for the fire to enter. Open up around the ceiling light fixture and then around any heat riser pipes or ducts in the area. Don't forget the range hood. Many range hoods aren't connected to ductwork, but you must check to be sure. Check for extension all along the run of the ducts. Does the duct pierce a wall or ceiling, or does it run hidden in the soffit? Did flames lap out of the duct's termination point? Did they ignite siding or reach through an open window above? Did they expose a roof overhang, allowing fire to enter the attic? You'll have to check.

You may have to move the stove to check behind it for extension. Take care when stretching the flexible gas connector, since it may develop a leak when you disturb it. If the burner or oven controls were damaged in the fire, you may have to shut off the gas to the stove. If the shutoff, typically found before the flexible connector, was involved in the fire or doesn't work, then shut off the gas at the meter.

Salvage Property. Remember, this is someone's home. Once you extinguish the fire, do as little damage as possible, and leave the entire home in as safe a condition as you can. Remove dishes from cabinets before removing the cabinets from the wall. Minimize the use of water while still making sure that nothing is smoldering. In short, pretend that it's your home and that you'll have to explain any damage to your spouse or mother.

MICROWAVE OVEN FIRES

Although microwave ovens heat without flames, they have been known to cause fires and generate smoke. A plastic container with a wire handle or a paper bag with a staple embedded in it can ignite when heated in a microwave. Aluminum foil can cause arcing and fire. When looking for the causes of an odor of smoke, don't forget to check the microwave. Even if the occupant claims not to have been using the microwave, check it. There have been reports of microwaves spontaneously turning on as a result of electric storms. In such a case, anything stored in there would be subject to heating and possibly fire.

SUMMARY

Kitchen fires range from being truly minor incidents to full-blown structure fires. Each requires different tactics and raises different concerns. These fires, like others discussed in this book, are often dismissed as being merely "routine." Don't make that mistake. Always take the appropriate steps to protect you, your brethren, and the occupants. Likewise, always be suspicious of possible fire spread. It's better to do a little too much damage than to leave smoldering fire behind when you pack up to go.

STUDY QUESTIONS

1. When responding to a food-on-the-stove call, the type of incident that you encounter depends in large degree on _____.

2. True or false: The method of forcible entry used at a food-on-the-stove incident should be in strict accordance with SOPs so as to protect the department against legal action.

3. If you receive a number of calls for an odor of smoke at a multifamily residence, need you check out each one?

4. Smoke from burning food probably contains nitrogen dioxide. Name some of the symptoms of NO_2 poisoning.

5. What is perhaps the best way to open a double-sash window to ventilate an apartment?

6. If a firefighter encounters heat while searching for victims and the source of the smoke, it is an indication that _____.

7. Is extension a possibility if you find a pot that is only smoking?

8. True or false: Spraying water into a pan of burning cooking oil may cause the oil to splatter, but it will suppress the flames.

9. Is a glass of water enough to put out a fire in an oven?

10. If the shutoff typically found before the flexible connector was involved in the fire or doesn't work, then you should _____.

Chapter Seven
Mattress Fires

After a night on the town, a man comes home a little drunk. He's hungry, so he fries up a steak, eats it, has one more beer, and then lies down on the bed to watch an old movie. He lights up his last cigarette. As he watches the film, his eyes get heavy, and his hand falls from his chest onto the bed. The cigarette drops from his hand and falls onto the blanket, where it continues to burn.

The blanket doesn't burst into flames. It smolders as the glowing embers burrow down through the blanket, the sheet, and into the mattress. As smoke builds up in the room, the man starts to snore. In the hallway outside of his opened bedroom door, the smoke alarm blares its warning. The man is startled awake. Sensing something wrong, he sits up in bed and breathes in a lungful of acrid smoke. Vaguely realizing the problem, he tries to get out of bed, but still feeling the effects of alcohol and now carbon monoxide, he's disoriented. He falls onto the floor next to the bed and doesn't get up.

An upstairs neighbor, hearing the alarm, wakes up and smells smoke. He has been through this nightmare before in a previous apartment building. He quickly calls the fire department, then heads for the door. There's no smoke in the hallway, only an odor. He rushes downstairs and exits the building anyway. He doesn't want to be in a burning building again.

By the time help arrives a few minutes later, smoke is pushing out of an open third-floor window. The firefighters spring in action, stretching a 1¾-inch line into the building and up the stairs. They force the apartment door open, and heavy smoke fills the once-clear hallway. The firefighters start their search.

When the officer finds the involved bedroom, he calls out to the other members of his team. While waiting for his can man to find him, he begins to search the room. The bed is in flames, but he tries to get as close to it as possible. He finds the victim and immediately announces the discovery over his radio. As he is pulling the victim out of the room, the irons man arrives and helps him. The can man gets close to the burning mattress, puts his finger over the nozzle to create a spray pattern, and empties the contents of the extinguisher into the flames, quelling them. The victim is removed to the floor below and is revived. He's one of the lucky ones.

When the engine arrives at the apartment door, the officer calls for water. The line is charged and moved into the apartment as far as the bedroom door. The fire is out and the victim has been removed. Can the firefighters just pack up and go home? Is the job over?

Not quite. The prime mission is to save life. So far, one life has been saved. Are there any others at risk?

AFTER THE KNOCKDOWN

Fire Department Tactics

Search. You must search the entire apartment for additional victims. Obviously, the search doesn't end when you find one. Any victim, when found, must be removed to a safe area and given medical attention, if necessary. In the above scenario, the man was removed to the floor below the fire, where the air was relatively clean and where firefighters could do what was necessary to revive him. They could have taken him to the street, but using the floor below allowed the firefighters to attend to his needs sooner. Oxygen and even a defibrillator could be brought upstairs, if necessary. If an ambulance is already on the scene, taking the victim downstairs might expedite advanced treatment, as well as transport.

Check Nearby Apartments. Check adjoining apartments and the apartment above. Smoke or fire may have extended there. Someone with a heart condition, asthma, or other pulmonary condition might be adversely affected by even a little smoke. In a recent fire in New York City, an elderly woman was found dead in a chair by an open window. This was unusual, since she wasn't in the fire apartment, nor was she above the fire. She was on the same floor as the fire, but across a courtyard from it. A search showed that the apartments next to hers were clear of smoke. Apparently, the smoke that entered her open window was enough to kill her. It would be easy to miss such a fire-related fatality. You must decide at each incident how far you should extend your search.

Check for Extension. Just because this was a mattress fire doesn't mean that the fire didn't extend. Depending on how long it burned, as well as the construction and condition of the building, the fire may have extended. For a smoldering mattress fire that didn't reach the active flaming stage, minimal overhaul is needed. The ceiling needn't be pulled if there was no active flaming. However, if flames and highly heated gases were produced, it's important to check to see whether the fire penetrated the ceiling or walls. The obvious areas to check are at ceiling light fixtures, pipes that penetrate the ceiling, electrical outlets and switches, and any other opening that might allow fire to extend behind the plaster.

Pull the Ceiling. An intact plaster or plasterboard ceiling or wall can effectively halt the spread of fire. However, if the ceiling has sustained damage from the heat, if it is cracked, or if parts of it have dropped, you must check it for fire. Open up examination holes at all likely avenues of extension. If you find charred wood, continue opening up the ceiling until clean wood is revealed. The end result of such an examination may be that you wind up pulling the entire ceiling, and plaster and lath cover the floor. Consequently, your search for life may be impeded by the debris, and you may have to move it to perform a thorough search. A victim can be covered by plasterboard and might not be found. One hidden victim was found a week later by workers who were hired to renovate the burned-out apartment. It makes sense to perform the search before all of the debris falls and to look under any debris that may be present.

Open Up the Walls. Is the wall too hot to hold your hand on? If so, open it up. Has the plasterboard fallen off in places, and is charred wood showing? You must open it up until clean wood shows. Fire extension low in a bay necessitates that the bay be opened near the ceiling. If you find evidence of fire there, you must open the ceiling and check the floor above for extension.

Fire can enter a wall through a wall switch or outlet. Check around them and open them up if you suspect extension. If the building is of balloon-frame construction, consider the possibility that any fire that penetrated the wall might have dropped to the floors below or extended to the floor above, even into the attic or cockloft.

Open Closets and Dressers. Both closets and dressers are combustible and are filled with flammable contents. They may be ignited by the heat of the mattress fire and may smolder for some time. The smoldering items not only hinder operations by obscuring vision and producing carbon monoxide, they may also ignite later, after the department has left the scene. You must open closets and drawers to make a thorough examination. Remove any burned or smoldering items, and thoroughly soak them with water. Use either a pail of water, the accumulated water on the floor, or a hoseline. Removing any burned articles from the building provides added assurance that there won't be a rekindle after you leave. At the same time, you mustn't needlessly damage salvageable items nor toss them into the street. Clothing, money, jewelry, and sentimental items can and should be saved.

Stretch and Charge a Handline. Thoroughly soak the mattress with the handline, even after dousing the fire with an extinguisher. Fire can burrow into a mattress and smolder there, protected from any water that you apply to the surface. If the mattress hasn't been cut open and thoroughly doused, it can reignite after the fire department leaves the scene. Opening up a mattress with a knife allows water to penetrate to the core, extinguishing any fire that may be hiding there. Some firefighters carry a folding, hooked-blade linoleum knife for this purpose.

144 • Responding to "Routine" Emergencies

Fire can burrow deep into a mattress and remain undetected for hours.

Foam mattresses or those with foam pads will continue to generate combustible gases, even after the active flaming has been extinguished by the 2½-gallon water extinguisher. These hot, combustible gases can collect around the mattress and even fill the room, looking for an ignition source and some oxygen-rich air. Venting a room at this time would supply oxygen to these gases, and in the presence of an ignition source, the entire room would be endangered. Thoroughly soaking the mattress will sufficiently cool it and stop the generation of these flammable gases.

Remove the Burned Mattress From the Building. Even after opening and wetting down the mattress, you might still have missed a smoldering ember in the stuffing. Get the mattress out of the building so that it can do no harm if it ignites later on. Taking it out, however, can be easier said than done, since mattresses tend to be bulky and hard to handle. Add the weight of the absorbed water, and removing it can be downright difficult. As firefighters wrestle it to a nearby door or window, the mattress will be bent this way and that, creating a bellows effect on any smoldering embers hidden within. It isn't unusual for a mattress to erupt into flames as it is being pushed out of a window. The bellows effect and the fresh air supplied to it can be enough to cause a flare-up. A mattress suddenly engulfed in flames can burn the unsuspecting firefighters or promote other injury as they try to jump clear of it. To avoid such an eruption, fold

Rather than extinguish this smoldering mattress inside the building, firefighters carry it outside, where it'll be cut open and soaked.

the mattress in half so that the burned side is within the fold. This will prevent oxygen from reaching the burned surface as well as the interior, thwarting a flare-up from the core. Tie the mattress in the folded position with a short length of rope. Tying it will also make it easier to handle.

You can then toss the folded, tied mattress out of a bedroom window that has been cleaned of glass and sash. Whenever you throw anything out of a window, first make sure that no one is standing below and that no victim has jumped and is lying below. If possible, assign a firefighter to inspect the area and keep everyone clear.

It isn't always possible or desirable to remove the mattress via the window. If it's on the upper floor of a high-rise, or if there are balconies, porch roofs, clotheslines, wires, or fire escapes below the window, it may be better to remove the mattress via the door. Doing so, however, entails some danger, for it's possible for the mattress to flare up while members are carrying it. A charged line must be available to protect them. Avoid using an elevator. Isolated in an elevator, you'd forfeit the protection of a handline. It's better to haul the mattress down the stairs, but provide enough assistance, since it'll be heavy (especially if it's wet) and awkward. If it's absolutely essential to use the elevator, take one or more extinguishers along with you, and keep a watchful eye on the mattress. At the

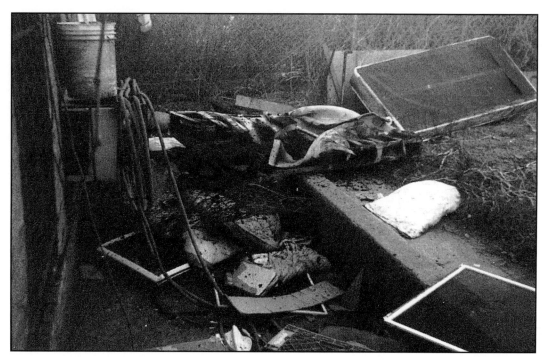

Before tossing a mattress (or any other overhaul material) out of a window, check the area below for any victim who may have jumped.

first sign of flame or smoldering fire, douse it immediately. Stop the elevator at the nearest floor to provide a means of egress should the extinguishers not be able to control the fire.

The same precautions that you would take for a mattress fire also generally apply to fires in stuffed furniture. The danger of a smoldering fire going unnoticed in an overstuffed chair is a very real threat. If the burned area is superficial and truly small, cut it out and soak it in a bucket. It'll still be necessary to open up the cushion and check the stuffing. If any doubts remain, soak it all down and remove it from the premises. You can soak the cushion in a tub or the sink, but you must take care not to clog the drain, since you may thereby cause a flood and unnecessary damage as a consequence.

SUMMARY

Mattress fires can result in civilian death or injury, and an aggressive search is needed to find and remove victims. The danger, however, isn't over once you make the rescue. Always bear in mind that there may be inobvious victims elsewhere in the apartment or in the surrounding units. Be cognizant of the potential dangers to firefighters from sudden flare-ups while moving the mattress, and consider the chance of a rekindle if the mattress isn't overhauled or removed from the premises.

STUDY QUESTIONS

1. After extinguishing a mattress fire in an apartment, where must you search?
2. Even after dousing a mattress fire with an extinguisher, you should _____.
3. To thoroughly extinguish a smoldering mattress, you should _____.
4. True or false: Extinguishing the fire in a foam mattress will curtail the generation of combustible gases.
5. To avoid an eruption of flames when pushing a mattress out a window, how should you prepare the mattress?
6. True or false: It is advisable to use an elevator when that is the fastest way to get a burned mattress out of the building.

Chapter Eight
Trash Fires

The call is for outside rubbish. A single engine is dispatched, and the reason for the call turns out to be burning rubbish in a wire-basket trash can. Rather than stretch a length of hose and then have to drain and repack it, one firefighter grabs the trash can with his hook and drags it over to the pumper. The pump operator charges the pump with booster water and cracks open a discharge gate on the street side of the pumper. Water flows out of the gate and into the garbage pail, now being tilted and held in place by the first firefighter. As the water flows into the trash can, quenching the flame, a loud pop is heard as a hot bottle shatters from being doused with cold water.

The firefighter drops the trash can and grasps his face, crying out in pain. A fragment of glass, projected from the shattering bottle, has embedded itself in his eye. He staggers in pain and confusion, spilling the smoldering rubbish into the street. The other firefighters help him onto the apparatus, and they speed off to the nearest hospital.

The above scenario isn't just a story. This type of injury has occurred more than once at rubbish fires. At incidents as insignificant as the one described, standard precautions are often overlooked, and injury is sometimes the result. The injury won't be incurred by an attempt to save life or even property, but rather, because personnel failed to wear the proper safety equipment and tried to avoid the trouble of repacking one length of hose or having to recharge a pressurized water extinguisher. A standard water extinguisher can project $2^1/_2$ gallons of water twenty feet or more, and it would certainly have kept the firefighter out of harm's way. He wouldn't have suffered an injury or risked blindness, the pumper wouldn't have been placed out of service for the trip to the hospital, and the department wouldn't have been socked with needless medical costs. All in all, laziness at a small incident can prove to be quite costly.

When you are called to a rubbish fire, in all likelihood, you won't know what's burning. The fire may be in tidbits of trash tossed from passing autos over a long period of time, or it may be in construction debris dumped at night by a contractor. The rubbish can be simple paper, leaves, scrap wood, or illegally discarded medical waste. In any case, all that you'll see will be the flames and the smoke coming from the pile. At night, identifying the contents of the fire will be even more difficult. You must at all times take precautions to protect yourself and others from whatever potentially hazardous materials are burning.

GENERIC RUBBISH FIRES

Smoke Hazard

What's in the smoke produced by a trash fire? You can't know, but rest assured, it isn't good. If rubber, carpeting, or upholstery is burning, hydrogen cyanide could be present. Acrolein is produced by burning carpeting. If the trash is plastic, hydrogen chloride could be in the smoke. If the plastic is PVC, then phosgene will also be produced. Is there a can of bug spray inside of a burning plastic garbage bag? Is a turpentine-soaked rag burning? Burning wood can give off formaldehyde. Is the wood saturated with creosote, a known carcinogen? Railroad ties are, and when they burn, the creosote is carried in the smoke. The

You never know what's burning at a rubbish fire, so stay out of the smoke and keep the wind at your back. Better still, wear SCBA.

list is nearly endless, and the question is, which and how many such health hazards are you willing to breathe into your lungs? Once you decide the answer to that question, ask yourself how you'll know which ones—and what quantities—are present at a particular trash fire.

As a new firefighter in training, I was required to remove my mask in a heavily charged smoke room, listen to a lecture, and answer questions until I thought my lungs would burst. The experience was unpleasant, but I thought that it had benefited me. Early in my career, it wasn't considered manly to wear a mask at a structure fire, much less at a rubbish fire, and I prided myself on being able to "take smoke." Well, my view of taking unnecessary smoke changed several years later when one-third of my lung was removed because of mysteriously damaged tissue. Suddenly, training in live smoke to "get used to it" no longer seemed attractive, and I have become a strong advocate of wearing masks. Don't enter or breathe any smoke that you can avoid. In recognition of the myriad dangers posed by smoke, fire departments now mandate the use of SCBA at structure fires. Smoke from a run-of-the-mill trash fire can be dangerous, too, even though it isn't confined. Either stay out of trash smoke or wear SCBA. Protect your lungs.

Traffic Hazard

Rubbish fires can occur on multilane highways, residential streets, and at any other location where there is vehicular traffic. As always, the prime directive is to save life. To accomplish this, a prime consideration at a roadway trash fire must be to protect firefighters from being struck by moving vehicles and to prevent civilian motorists from crashing into fire apparatus. If the location of the fire requires that firefighters operate on the roadway, position the apparatus between them and the oncoming flow of traffic. If firefighters must remove or replace hose or tools at the rear or traffic side of the apparatus, position a second vehicle as a shield to the first. At all roadway incidents, the second unit should take this position as a safety precaution. You should also place flares as described in Chapter Five. If the fire is located on a blind curve or just beyond the crest of a hill, or if traffic is moving swiftly, consider placing flares ahead of the scene to warn oncoming drivers. If available, use the police to control the flow of traffic. In any case, complete the job as quickly as possible, and get your firefighters out of harm's way.

Rubbish Fire Tactics

Remember, this is a trash fire. There is no civilian life hazard—you are the only life hazard. There is no property to save—whatever is burning has been discarded and isn't worth saving. You must, of course, prevent the fire from extending to brush or a structure and becoming a bigger incident than it is.

Stay Out of the Smoke. Using SCBA at every rubbish fire will protect you from the smoke generated by burning rubbish, but in the real world, it isn't likely that you'll use a mask each time. You can, however, in many cases stay out of the smoke by using the reach of your line and attacking the fire with the wind at your back. The wind, of course, may be variable, and you may have to reposition each time it shifts.

Select an Adequate Hoseline. Whichever line you choose must have enough reach to keep you away from the rubbish while delivering enough water to quickly extinguish the flames. For a small fire, a 2½-gallon water extinguisher may do the job. A booster or trash line may be adequate for some fires, but for larger fires or more hazardous ones, a 1¾- or 2½-line may be required. For a large, deep-seated fire, such as you might encounter at a garbage dump, an aerial platform stream might be the best choice. If you opt for the elevated stream, remember to keep the firefighters in the platform either out of the smoke or suited up in SCBA.

Avoid Direct Overhaul. If the burning rubbish covers a large area, try to avoid manual overhaul, which may necessitate that you enter the smoke and walk into the rubbish. You would risk a puncture injury to your feet or legs, a sprain, or a fall, not to mention the chance of being hit by a projectile from an exploding container. If the scene turns out to be a haz mat incident, you will have contaminated all of your gear. Consider using your deck gun or an aerial stream to overhaul a large area or a high pile of rubbish hydraulically.

DUMPSTER FIRES

Large trash dumpsters are found at construction sites, commercial occupancies, buildings under renovation, and anywhere that large amounts of disposable rubbish are generated. There's no telling what a metal dumpster might contain. It may be intended for construction trash at a demolition site, but passersby might illegally dump their own refuse into it, including hazardous materials of any description. Asbestos, PCBs, various chemicals, biological waste, and anything else that you can think of might be inside a burning dumpster. It's common to find broken glass, razorlike shards of metal, and nail-riddled boards in such containers. The type of occupancy served by a given dumpster can give you a hint as to what you might find. Is it behind a restaurant, a clothing store, or a building under renovation? If it serves a hospital, you might find needles, illegally dumped red-bag bio waste, and other hazards. Moreover, the smoke generated by the burning trash can obscure the color of the bag until after the fire has been extinguished. By then you will already have been exposed. Although the type of occupancy gives you a clue as to what to expect, any dumpster is prone to illegal dumping of hazardous materials.

Dumpster Fire Tactics

One effective method of extinguishing such a fire is by using the deck gun of a pumper to flood the container. Position the apparatus so that the stream can

There's no telling what a dumpster might contain.

sweep the dumpster. You'll have to stretch a handline if this isn't possible or if it doesn't effect total extinguishment.

Precautions at a Dumpster Fire

Do Not Enter the Dumpster. Hazardous materials, sharp objects, and pressurized containers that might explode should be enough to keep you out of a dumpster. In addition, the chance that a fire is burning within a void means that the surface might give way under a firefighter's weight, possibly causing a fall injury. Dumpsters can be crammed full of heavy, tightly packed trash, and they are inherently dangerous places.

When Possible, Overhaul the Contents Hydraulically. Attempting to overhaul the contents manually can result in strain and sprain injuries to firefighters. Use your hose stream to overhaul the contents of the container. Large dumpsters often have a door on one or more sides that can provide access for a hose stream. On smaller dumpsters, a hinged metal cover is the norm. If necessary, you can push and shove the contents with a tool, but doing so puts you close to the container and the hazards within.

Remember That You Are the Life Hazard. No matter how you decide to extinguish the fire, preventing injuries to firefighters must be the primary concern. It's unlikely, yet possible, that there is a life hazard in the dumpster. A vagrant might have been foraging for salvageable items or even sleeping in it, or children might have been playing in it and become trapped or injured. This isn't to suggest that you should conduct a search before applying water—just bear in mind that there's always a remote chance that someone is inside. Look for signs of occupancy, such as clothing or utensils. A spectator may tell you that a child or vagrant has been seen in the area recently. If you are told of a possible victim, you may want to restrict the use of water, and you'll have to make a thorough search as soon as safety permits.

Consider Exposure Problems. Dumpsters are often positioned near a building. A flaming fire could extend, especially if the siding of the structure is flammable or if the dumpster is below a window. Position your handline so that it can wet down the exposure and apply water to the fire. You may have to conduct an examination of the building's interior if extension is a possibility. It may also be necessary to remove some of the combustible siding to check for smoldering embers in the insulation. Another possibility is that the flames have exposed the eaves of the building. In this case, you'd have to check for extension in the attic. If extension into the structure is possible, call for help early. Playing catch-up with a structure fire is rarely successful. In some cases, a single engine can safely extinguish a dumpster fire, but if it has extended, you'll be dealing with a structure fire. The engine officer in command should call for a structural alarm assignment, and he should make this call sooner rather than later. Don't wait until you force entry and discover that you have a structure fire on your hands. One way to prevent extension from smaller dumpsters is simply to roll them away from an exposed building. In most cases, this can be done by one or two members. Once you move the dumpster, the danger of extension is over and the need for additional units removed.

JUNKYARD FIRES

In my district, there are several junkyards containing automobiles in various stages of decay and disassembly, often piled one on top of another. By law, the gas tanks should have been drained or removed, but you never know. A gas tank may be just a chunk of metal that has been compacted in a crusher, or it may still be intact. There are piles of tires, racks of auto parts, acetylene and oxygen tanks for dissecting autos, tanks of flammable liquids, and the shredded remains of upholstery and other nonmetallic car parts.

Typically, a fire in a junkyard is set by the careless use of a torch, a discarded cigarette, or by vandals. If the employees accidentally start a fire while cutting apart a car, they'll usually try to extinguish it themselves. It's only when they fail and the fire spreads that they call the fire department.

At this junkyard incident, gasoline leaked from a failing gas tank and touched off a spate of vehicle fires. Luckily, the firefighters could stand back and use the reach of the stream to douse them. (Photo by Matt Daly.)

Tactics at a Junkyard Fire

Recently I responded to a junkyard fire that turned out to be an old twenty-two-foot center-console boat, fully involved on our arrival. It was generating large quantities of thick black smoke as its fiberglass hull burned. A number of questions instantly came to mind. Was there any gasoline in its tank, or were there gasoline vapors trapped in its hull? Were the batteries still on board, or hazardous materials? Were any people on board? The immediate answer to all of them was "I don't know."

Had anyone been on board, he would have been dead. There were no exposure problems. If left alone, the fire would have burned itself out. A slow, cautious approach was indicated. The first pumper emptied its booster tank through its deck gun. This didn't kill the fire, but it did darken it down. Meanwhile, the second engine hooked up to a hydrant to stretch a 2½-inch hose, which was able to extinguish the remaining fire from a safe distance. The

156 • Responding to "Routine" Emergencies

At this nighttime boat fire in a junkyard, firefighters were the only life hazard. Still, you can't always guarantee that such vessels and vehicles will be unoccupied.

truck team conducted minimal overhaul, and we were able to take up without incurring any injuries.

Determine Whether There Is a Life Hazard. Question the employees. Find out what happened and whether anyone was injured or is missing. Find out what's burning, whether it's threatening a nearby structure, and whether a life is at risk in that structure. If the fire is on or near a storage rack, consider the possibility that the rack could fail, dropping its contents. If the fire occurs in the middle of the night as a result of vandalism, there will likely be no one to question, so be cautious.

Select the Appropriate Line. For a large body of fire, the best choice may be the elevated stream of an aerial platform. The large-caliber stream can be used to extinguish fire from a variety of angles, and it can be used to drive a stream into the top of a pile of burning material without having to place a member on the pile. The appropriate line will be the one that gives you ample reach, as well as a flow that will quickly extinguish the flames. You must weigh the mobility of a $1^3/_4$-inch line against the reach and flow of a $2^1/_2$-inch line. Remember, you can

An aerial platform moves in to extinguish and overhaul this deep-seated junkyard fire.

always add a length of 1¾-inch hose to the larger one once the fire has been knocked down. That way, you'll have the initial benefit of longer reach and heavier flow, and when the time comes to change position, the smaller line will be easier to haul around. You must also consider the water supply, since any hose is valueless unless the water supply is capable of delivering the proper flow.

Minimize Overhaul. Overhauling auto parts may necessitate being in close proximity to storage racks or piles of cars that could collapse. Use a minimum of staffing, but don't allow one firefighter to move heavy objects by himself. Either perform the overhaul hydraulically, or employ several firefighters to do the heavy work.

When Available, Use Heavy Equipment. Junkyards routinely use backloaders, high-lows, and cranes to move wrecked autos around. Without heavy equipment, it may not be possible to effect complete extinguishment of piled auto parts or other refuse. A burrowing fire in a large pile of granular material may continue to burn even after extended attempts at extinguishment. Such a pile would have to be pulled apart and wetted down, but without the appropriate equipment, this can't be done. In instances when I need heavy digging equipment

Expect these light metal storage racks to fail early if a fire ever occurs at this site. Anyone working on or near them could be crushed as stacked cars and auto parts begin to topple.

At this incident, a crane was used to pick burning cars off the pile and set them down on the ground within easy reach of a handline.

but none is available, such as in the middle of the night, I do the best I can with hydraulic overhauling and then cover the area with protein foam. The foam has proved effective in preventing a flare-up of the burning material, at least until the next day when the required equipment becomes available.

At a fire in a high pile of junk autos, after the bulk of the fire was knocked down by an aerial platform stream, it was necessary to use a crane to lift cars off the top of the pile and place them on the ground, where final extinguishment could be effected by a handline. The steaming auto hulks were then piled up at a nearby location into a new pile. The burning pile was thus systematically taken apart, extinguished, and relocated.

STRUCTURAL RUBBISH FIRES

Vacant buildings, unless properly sealed, can become large garbage pails. Neighbors, children, contractors, and passersby have all been known to dump their trash into unsealed vacant buildings. Add to this the chance of kids playing or vagrants living in there, and you have the potential for a dangerous fire. All of the hazards of trash fires will be present, plus the dangers of structure fires, including collapse. Such a building, if left for a prolonged period of time, will suffer structural damage from the elements. Beams, floors, and roofs will rot. Repeated fires will further weaken the structure, and accumulated trash will create a heavy fire load, absorb water, and add weight to the floors.

Look for signs of squatters or children. The presence of clothes, bedding, or food, or perhaps an accessible opening in an otherwise sealed building, may indicate that it's inhabited. It's also possible that the building is a drug den. Contaminated needles or booby traps might be present. A fire in such a structure might be a haz mat incident as well as a structure fire. Serious potential for firefighter injury exists in such buildings.

Tactics at Structural Rubbish Fires

Beware of the Dangers. Consider all such fires to be extremely hazardous, even if the fire is small. Holes in the floor, ceiling, and walls; missing doors; and a variety of rubbish strewn about will allow fire to spread with alarming speed. Entering a vacant building under ideal conditions can be dangerous enough. Add to it reduced visibility, and the risks only increase.

Evaluate the Building Before Entering It. Past knowledge of the structure will help you decide whether to enter or to conduct an exterior attack. Has it been the site of numerous fires? Has it been opened to the elements for a long time? If the structure has been seriously weakened, it may not be prudent to place firefighters inside except for a truly minor fire.

When Possible, Extinguish the Fire From the Outside. Use the reach of your hose stream to keep you out of the building. If the fire is in a room that is accessible, extinguish it from the exterior. Then, if safety permits, enter the building to complete extinguishment and perform overhaul. Use the minimum amount of personnel that will get the job done safely, and use portable lights to illuminate the area. If you conduct interior operations, shine lights up through holes in the floor and stairways to prevent firefighters from walking into them. Holes in the floor can be made safe by covering them with doors taken from adjoining rooms. Even if you decide to conduct the attack from within, you should still let the reach of your stream work for you. Extinguish as much fire as possible from the doorway before entering a given room. Then, cautiously examine the room before you step inside. Using extra caution will help you avoid holes in the floor and other hazards. Often you can easily spot holes from the floor below the fire. Assign someone to check out the floor below before you move in on the fire.

Search for Extension and Other Fires. It's possible that more than one fire has been set, or that fire has extended to the floor above or below. You must check for extension, not only to prevent a rekindle, but also to protect firefighters from becoming entrapped by extending fire. When searching such a building, firefighters, preferably in pairs, should be equipped with a radio.

If firefighters are operating abovegrade, raise ladders to provide them with an alternate means of egress. If the windows are sealed, open them for ventilation and as a ready escape route.

Unlike a pile of rubbish burning on the roadside, there may be dire consequences if you leave a vacant-building fire smoldering. If it were to reignite, it would burn not only the rubbish, but also the structure and its contents. It could also spread to adjoining structures, and in so doing, endanger life. When deemed safe, enter the building with masks and full firefighting gear. Maximum supervision and a minimum commitment of personnel can reduce the possibility of firefighter injury.

It may be possible to use exterior streams to extinguish as much of the fire as possible before entering the building. After the bulk of the fire has been extinguished, and before committing firefighters to interior operations, a senior chief or officer should reevaluate the building to determine whether it's safe to enter. If so, position a rapid intervention team (RIT) at the ready to move in should the interior personnel become injured or trapped. Firefighting shouldn't be conducted in any derelict building without adequate resources standing by to assist in an emergency.

If the building is too hazardous for entry, hydraulically overhaul it with a large-caliber stream. An aerial platform stream is ideal for this task. It can go from window to window, opening up walls and ceilings with the force of its stream, placing water just where you need it. Realize, too, that it can also cause collapse, because it delivers so much water with so much force. If the water is absorbed by the plaster or accumulated rubbish, or if the water pools on the floor, collapse potential increases. While using elevated streams, the incident commander should also be looking for runoff. Since an aerial platform can deliver as much as 1,000 gpm, water should be pouring out of the building. If it isn't, then it might be collecting on the floors or being absorbed, adding tremendous weight to the weakened structure.

An aerial platform stream can also blow shingles and roof boards off the building, and even knock bricks from the walls. The force of such a stream alone hitting a wall or floor could precipitate a major collapse. During operations, a collapse zone should be established and enforced until such time as the incident commander decides to initiate interior operations. Remember that the aerial platform must be kept out of the collapse zone as well.

Mark the Building. You should mark the building so that firefighters who respond to future fires there will be aware of its structural condition. We regularly mark our vacant buildings using lime yellow spray paint and one of three symbols. For a building with normal stability, we paint an 18-inch by 18-inch square over the entrance door to indicate that, at least at the time of marking, no unusual hazards were present in the building. If structural damage or interior hazards are present, we paint an additional line within the square, going on a diagonal from the upper right-hand corner to the lower left-hand corner. This means that an examination must be made before interior operations can be conducted. If the decision is made to operate inside, extreme caution is required.

A square with a full X painted in it means that severe interior hazards or structural damage exists and that all operations should be conducted from the exterior. If an interior attack is contemplated, the officer in command must approve it, an interior survey must be conducted, and the operations must be carried out by a minimum number of personnel. Frequently, fires in buildings marked this way are extinguished by streams from aerial platforms and are overhauled hydraulically.

EXPOSURE FIRES

Occasionally we respond to a rubbish fire that is near to or up against a building. The rubbish fire presents no problem to us and is easily extinguished. However, another question presents itself: Did the fire extend into the building? This is a question that must be answered quickly. Sometimes determining the answer is a simple matter of looking through a window to see whether fire has extended, but often it's not so simple.

Tactics for Exposure Fires

Effect Timely Extinguishment. When confronted with a rubbish fire against a building, you should extinguish the fire as quickly as possible. The sooner you do so, the less likely it is that extension will be a problem. Choose a hose size that'll be able to deal quickly with the volume of fire, yet one that you'll also be able to haul into the building quickly if the fire does extend. If you suspect extension, stretch a second, precautionary hoseline to the interior as soon as possible, even before you've extinguished the rubbish fire. If life is in immediate danger as a result of the rubbish fire, place the first line to protect life and stretch additional lines to extinguish the rubbish fire plus any extension.

Suspect Extension. Check the exterior wall closely for cracks, holes, or other openings that might allow the passage of fire. One small opening is enough to promote the ignition of an interior wood framing member. Such extension won't be noticeable looking through a window from the outside. Check to see whether there are overhanging eaves, and determine whether the flames were high enough to reach them. Was the heat intense enough to ignite the underside of the eaves, or did sparks rise and become trapped in cracks there? Such a minor transference could, if undetected, translate into full-fledged extension.

The examination for extension can start on the outside. Look for cracked windows, or signs of fire or smoke through a window. Perhaps smoke is pushing or seeping out from under the siding or through a crack in the wall. If there is any chance that fire has extended, you must enter the building to conduct a survey. Although this may be no problem if the occupant is on hand, you may hesitate to force entry and cause damage to an unoccupied building just for the sake of an outside rubbish fire. Still, whatever damage you do to the building in gaining entry will be small compared with what hidden fire could do.

If the fire was truly a small rubbish fire, it's less likely to have extended. The risk increases with intensity. Was the fire burning against a brick wall? If so, direct extension is unlikely, but embers may have risen to a window, the eaves, or the roof. Is fire breaking out on a bed underneath an opened window, or is fire extending to the attic through the roof? Was the fire burning against aluminum siding? Was the underlying framing or insulation ignited as the heat of the flames was conducted through the metal siding? You'll have to remove some of the siding to check. If it's possible that fire has extended into the building, you may have to open ceilings and walls.

If you discover fire, quickly open up around it to check for further extension. Ask yourself whether the building is of balloon-frame construction. If so, check the attic and cellar, and apply water to the areas where fire has extended. This may be done with a water extinguisher or a handline, depending on how much fire is present. Use as much water as necessary, but don't cause unnecessary damage to the occupancy. This is especially important in the case of someone's home. Flooding a basement isn't necessary for ten cents' worth of fire, and it won't win you any popularity with the homeowner. Use as much water as you feel is necessary, but be ready to explain your actions to an irate occupant.

SUMMARY

Trash fires are perhaps the easiest fires to take for granted. We respond to so many of them that we frequently fail to view them as threatening, yet they are regularly the cause of painful, crippling injuries. By routinely taking basic precautions, firefighters can cut down on such injuries. Still, it is up to the officer, who must take the lead, insisting that all members wear proper protective gear and follow safe firefighting practices.

STUDY QUESTIONS

1. What is the best way to stay out of the smoke if, for whatever reason, you aren't equipped with SCBA at a rubbish fire?

2. Why isn't the type of occupancy an adequate indication of the hazards associated with a fire in the dumpster out back?

3. True or false: Because of the many void spaces in trash, it is better to overhaul a dumpster fire manually than hydraulically.

4. The primary life hazard at a dumpster fire is that of _____ .

5. Perhaps the best way to overhaul a pile of automobiles at a junkyard fire is to use _____ in conjunction with hydraulic overhaul techniques.

6. The author mentions a simple way of identifying holes in the floors and stairways of derelict buildings. What is it?

7. When directing a master stream into a derelict building, how can an incident commander judge whether too much water is collecting in or being absorbed by the structure?

8. What is the purpose of marking a derelict building?

Part Two
Carbon Monoxide

Chapter Nine
The New Response

In 1994, fire departments all over the country were struggling with a new type of response: the home carbon monoxide alarm. Personnel had little experience with these alarms, and virtually no one knew how they worked. In fact, many departments initially treated them just like smoke alarms. Called to a home, responders would often enter and, finding no apparent problem, declare the incident to be the result of a defective alarm, then simply leave. Over time, they realized that there was more to this type of alarm than was apparent at first glance. My experience with them was similar, as evidenced by the following incident.

We responded to a call for a smoke detector sounding in a private home. It was not an uncommon occurrence. We regularly went to two or three of these a tour. Mostly they were unwarranted or the result of friendly fire, such as cooking or smoking. Often the units would be defective. This one, however, turned out to be a CO detector. The truck officer, already inside the building, contacted me on his portable radio. He said, "Chief, it's a carbon monoxide detector, and it won't stop alarming. There doesn't seem to be a problem in the house. It must be defective."

I instructed him to bring the detector outside and to wave it in the fresh air to see whether it would reset. When it didn't, I told him to remove the battery. He said he couldn't find it, but then, after a brief pause, told me he thought he'd found it, and he brought the unit over. This was the first time that I had seen a home carbon monoxide detector.

Removing the battery pack stopped the alarm. I had the officer replace the battery pack to see whether removing it had reset the detector. The alarm started to sound again right there in the street.

"You're right," I agreed. "It must be defective." Uncomfortable with that decision, I directed the officer to check the occupants for symptoms of CO poisoning. They exhibited none. We then told the homeowner that the unit was defective, and we left the scene.

Fortunately, that was the end of that incident. I say fortunately, because at the time, we were all woefully ignorant of CO detectors and how they worked, as well as of the dangers of carbon monoxide in the home. Oh sure, we knew all about carbon monoxide dangers at structure fires. We had seen, read, or experienced the effects of breathing too much CO-filled smoke, as well as the dangers of backdraft, the explosive ignition of high concentrations of confined CO gas. We knew that carbon monoxide has over two hundred times more of an affinity for hemoglobin as oxygen, and that, in high concentrations, it can drop a firefighter in two or three breaths. As we responded to more and more of these alarms, and as carbon monoxide found its way more and more into the news, we would learn much more about the topic.

After responding to several other similar alarms and hearing about many more, I began to doubt that each of these incidents was in fact the result of a defective alarm. My department was in the process of formulating a response protocol for carbon monoxide alarms, but this offered me no immediate help. I needed to know more about this problem immediately, so I began to research it.

The first place I looked was in my copy of the *Fire Prevention Handbook, Seventeenth Edition,* which gave me basic chemical information on carbon monoxide, relating it to firefighting. This shed no light on my new concern regarding CO in the home and CO alarms. All of the other books and magazines in my firefighting library yielded similar results.

Next I searched through several medical reference books, learning more about the insidious nature of the chemical. This whetted my appetite for information, but I still hadn't found what I needed: a game plan to use at my next CO detector response.

Continuing my search, I ventured onto the information highway and cruised through several medical databases, looking for articles containing the words *carbon monoxide* and *poisoning* or *death.* One search alone turned up sixty related articles. Obviously, this issue was much larger than I had suspected.

I began calling people whom I knew in the fire and occupational safety fields, asking them what information they had on the topic. One lead put me in touch with the International Association of Fire Chiefs (IAFC), Operation Life Safety. This wing of the IAFC was dedicated, among other things, to promoting the use of both smoke detectors and sprinklers in the home. The IAFC was aware of the CO problem early on and, under sponsorship by First Alert, was working on a response protocol for the fire service. Apparently I wasn't the only fire officer who had been confounded by this new type of response. As luck would have it, the IAFC was planning a workshop in Kansas City entitled *Fire Department Response to Carbon Monoxide Detector Calls,* and I was invited to join in.

During the workshop, I learned that there were many aspects of the CO problem that I hadn't even considered, and that other departments had struggled with this topic already and had come up with some solutions and lots of questions. I came away from the workshop with a lot of information and a need to know more about the subject.

Since then, I have continued to research the various aspects of the problem. I and the other attendees of this workshop have since been charged by the IAFC to conduct seminars for fire departments on carbon monoxide response. In my department, the Fire Department of New York, I have participated in a pilot program, field testing various CO detection instruments, and I have lectured on the topic for various fire departments in the northeastern United States and Canada, as well as for the New York City Fire Department Institute and the Fire Department Instructors Conference (FDIC).

THE PLAYERS

Many different people and organizations are interested in the carbon monoxide problem, and all of them have their own agenda, including the fire service. Certainly the manufacturers of CO alarms have a vested interest in the issue. The more the general population becomes aware of the problem, and the more dangerous they perceive it to be, the more detectors they will purchase. The potential market for these CO detectors is estimated to be as high as two hundred million units. As one would expect, the manufacturers try to show their products in as favorable a light as possible. The home detectors sold today are being improved and features are being added in the hopes of enticing customers to buy. New technologies are being tested as manufacturers compete for a bigger share of the marketplace.

The manufacturers of gas detection instruments used by firefighters are also looking to maximize their profits on sales. As fire departments respond to this type of incident more and more, it stands to reason that CO meters will increasingly be in demand. A quick perusal of the various firefighting magazines will show you how aggressively these instruments are being marketed.

The gas utilities, propane suppliers, and oil companies are all keenly interested in this issue. Their products—natural gas, propane, and oil—are all sources of carbon monoxide in the home. As the number of CO detectors in American homes increases, gas and oil companies are responding to these calls with escalating frequency, and each such response costs the company money. If the fire service can screen carbon monoxide calls for these companies, then the companies stand to save a tidy sum. In New Jersey, one gas utility offered to supply the state's fire departments with gas detection instruments if the various fire departments would take on more responsibility for responding to these alarms. The initial cost of the instruments would easily have been offset by the savings realized by the reduced gas company response.

Along with the public's increased awareness of the dangers of carbon monoxide will come a demand for appliances that produce less of it. This represents an added expense for manufacturers as they try to develop cleaner-burning appliances. They, too, are watching developments carefully, and like all of the other interested parties, they're trying to favorably influence unfolding events.

The medical community is also concerned with this problem. Carbon monoxide poisoning is more pervasive than was previously thought, and the medical community's ability to diagnose it and treat it effectively can affect not only the patient's well-being, but also the medical professional's liability.

Let us not forget that fire departments, both career and volunteer, also have an interest in the outcome of technical and legislative developments regarding carbon monoxide. If laws are passed requiring CO detectors in homes (laws have been passed in Chicago and Kingston, New York, and an attempt was made to pass them in New York City and New Jersey), then fire department responses to such calls will increase. How often a department must respond to such incidents can affect its ability to deliver fire protection. On the legal side of the issue, what the department is responsible for at such incidents and what it's able to do can affect the department's, as well as the chief's, liability.

All of the groups mentioned stand to gain or lose time, money, or prestige as a result of the rapidly changing carbon monoxide landscape. Meanwhile, government agencies such as the Environmental Protection Agency (EPA) and the Occupational Safety and Health Administration (OSHA) are setting and revising air quality standards. These standards need to be monitored and enforced. As the standards change, the rules change, and all of the above groups are affected. The Consumer Product Safety Commission (CPSC) advocates having a CO detector in every home. Underwriters Laboratories (UL) has set the minimums for detectors in its standard UL 2034, and it has since changed this standard in response to information, data, and pressure from the above-listed groups and agencies, as well as from others.

Carbon monoxide detection and response are still in a state of change. The International Approval Services (IAS), a joint venture of the American Gas Association (AGA) and the Canadian Gas Association (CGA), has developed supplemental requirements to UL 2034. The stated purpose of these requirements is to reduce low-level activations of existing alarms and to improve the reliability of new home detectors. The IAS feels that, if new detectors adhere to these updated requirements, the resultant product will be greatly improved. The AGA's new residential *Carbon Monoxide Detector Standard of Design Certification, IAS U.S. Requirement No. 696,* became effective on October 3, 1996. There are no guarantees as to how this issue will play out or what the final rules and standards will be. One thing is certain: If you don't make the effort to learn the changing technology and keep informed about the changing regulations, your response to these emergencies will be inadequate at best, and your liability for improper action will be enormous.

In the following, I will attempt to give you a history of the carbon monoxide issue, as well as a snapshot of the situation today, while explaining how fire departments have responded to it. The information given isn't simple to digest, and it's best taken in small doses. Hopefully, it'll give you some insight as to what you must do to serve both yourself and the public well.

THE CARBON MONOXIDE PROBLEM

Of course, carbon monoxide isn't a new phenomenon. It's no more lethal today than it was when a caveman first carried a burning log into his den for heat and to cook his food. Why, then, is carbon monoxide so much in the news today?

That caveman's home was a drafty place with a large opening that readily allowed the free exchange of air. Carbon monoxide and other pollutants could exit the cave while fresh air entered. Even as recently as the late 1960s, homes were still relatively drafty, allowing an adequate air exchange. Unfortunately, this is no longer the case. Today, we build our homes tighter and more energy efficient. In the past, our homes were able to breathe, and there was an exchange of air from within the home to the outside environment. Pollutants could escape from within the home, and clean air was able to infiltrate inward. Now, improved building technology and an increase in the price of crude oil have changed all that.

In the 1970s, the price of home heating oil skyrocketed. As a result, we did all that we could to reduce our consumption of this precious commodity. We insulated our homes, installed energy-efficient windows, and covered other existing windows with storm windows. We put weather stripping around our doors and caulked all the existing cracks and crevices. In effect, we sealed in the normal household pollutants while locking out fresh air. This lack of fresh air can result in excess production of carbon monoxide by fuel-burning appliances and an accumulation of CO and other pollutants in the home. As a result, a CO problem was created in many homes.

Until recently, there wasn't much that could be done about the carbon monoxide problem. In fact, it wasn't even recognized as a problem. People were made sick and some died, but often the deaths and illnesses weren't attributed to carbon monoxide. CO is a colorless, odorless, tasteless toxic gas. That we can't smell it or taste it makes it insidious. We can breath it, be made sick by it, even be in danger of dying from it and never be aware. If we aren't aware of it, then we can do nothing to protect ourselves from it, nor will we attempt to remove ourselves from the danger. Still, if the exposure is continued and severe enough, we can die from it. This is just what was happening to homeowners all over the country, to the tune of 1,500 deaths and more than 10,000 illnesses a year, all as a result of accidental carbon monoxide poisoning. In fact, CO is the leading cause of accidental poisoning death in the United States.

A SOLUTION

A carbon monoxide poisoning incident in California that was caused by a faulty furnace resulted in the death of an entire family. Dr. Mark Goldstein was asked to investigate the incident. Dr. Goldstein, realizing that there was no affordable warning device available to a homeowner, began working on a solution, and the result was an affordable battery-operated CO detector that could warn occupants about a buildup of the deadly gas.

Since the first affordable home CO detector was created by Dr. Goldstein in conjunction with the Quantum Group, other technologies have become available to compete against it in the home market. It would appear as if the carbon monoxide problem has been solved. Not so.

These detectors, while effectively warning of a buildup of CO gas, gave fire departments and utility companies a whole new reason to respond to millions of homes around the country. Many firefighters, myself included, had no idea how to respond properly to a sounding CO alarm. Unfortunately, the errors that I made on my first response were the norm rather than the exception. Firefighters, it seemed, needed to go back to school to learn how to investigate such an alarm and to mitigate the hazard of carbon monoxide buildup. Few understood the technology of these detectors, and many didn't have the appropriate gas detection instruments required. Even today, unequipped and untrained firefighters are quick to declare an incident to be the result of a defective alarm and then simply take up, leaving a potential CO problem uninvestigated. Now, many departments carry gas detection instruments and routinely perform CO investigations, but often they are done in a cursory manner, if not incorrectly. The big loser is the occupant. If the responding firefighters don't have the right attitude, the essential knowledge, or the required tools, a potentially dangerous situation may go uncorrected, and illness or death may result.

SUMMARY

Responding to calls related to carbon monoxide detectors poses new challenges to the fire service, but it also provides new opportunities to serve the public. As new legislation seeks to better protect the citizenry, various agencies try to influence this legislation to their advantage. This changing landscape means that firefighters must keep up to date, not only with CO detector technology, but also with the laws and regulations pertinent to the industry. Without state-of-the-art detection equipment, proper training, and a working knowledge of the dangers posed by carbon monoxide, we won't be able to ensure the safety of those who look to us for help.

STUDY QUESTIONS

1. Carbon monoxide has more than _____ times the affinity for hemoglobin as oxygen.

2. True or false: Although burning oil and propane are well known to produce carbon monoxide, natural gas does not.

3. Underwriters Laboratories has set the minimums for CO detectors in its standard known as _____.

4. Why wasn't CO poisoning in the home as much of a recognized problem prior to the 1970s as it is today? Give two reasons.

5. By current statistics, how many deaths in the United States are attributable to carbon monoxide poisoning each year? How many illnesses?

Chapter Ten
The Medical Aspects of Carbon Monoxide

Too often, CO poisoning goes unrecognized. It's essential that you understand how carbon monoxide poisoning works, its symptoms, and how to treat it. If we as first responders aren't attuned to the signs and symptoms of CO poisoning, its victims may never be correctly diagnosed and treated.

Virtually any flame used in the home can generate carbon monoxide. It is even generated naturally by the human body, albeit not at harmful levels. At high levels, it can sicken and even kill its victims.

THE POISONING PROCESS

As a person inhales air, oxygen transfers from the lungs into the blood, where it attaches to the hemoglobin, which in turn transports the oxygen throughout the body. The hemoglobin freely releases the oxygen to the billions of cells, each of which needs O_2 to survive.

A hemoglobin molecule has four binding sites for oxygen. If carbon monoxide attaches to one of these binding sites, it acts in two ways to thwart the cellular respiration process. First, it takes up space on the hemoglobin that should be reserved for O_2. Second, it prevents the O_2 present from being released by the hemoglobin. If one of the binding sites is occupied by CO, the O_2 on the other three sites cannot be easily released. This dual effect results in tissue hypoxia, or oxygen starvation of the cells, thus suffocating the body from within.

This problem is compounded by the nature of hemoglobin itself, which has a greater affinity for carbon monoxide than it does for oxygen—between 200 and 270 times greater. This means that carbon monoxide is over 200 times more likely to attach to the hemoglobin than oxygen is, and that one part of carbon monoxide in the blood requires more than 200 parts of oxygen to replace it. As a result, even small amounts of carbon monoxide in the air can become fatal.

As one breathes carbon monoxide, and as the body is deprived of life-giving oxygen, the body tries to compensate by increasing its breathing rate. If the person remains in the contaminated atmosphere, his CO intake will be increased, and the body will respond by increasing the breathing rate even more. A vicious cycle is thus set up that, if not interrupted, can end in death.

Although most inhaled carbon monoxide is absorbed by the blood, about 15 percent is absorbed into body tissue. This tissue holds onto CO longer than the blood does, and this residual CO can be released from the tissue into the system after the patient has been treated, tested, and shown to be free of carbon monoxide. In this way, a victim of CO poisoning can suffer a rebound effect after being treated for it.

Factors Affecting the Severity of Poisoning

Carbon monoxide doesn't affect everyone the same way, and there are a number of factors that can increase its effect on individuals. Different people can exhibit different symptoms when exposed to the same amount of CO for the same duration. In one case, two people were found unconscious, suffering from CO poisoning. By the time they were brought into the hospital, one was sitting up talking while the other was still unconscious. It has been reported by the Maryland Institute for Emergency Medical Services that its hyperbaric facility has received patients with blood carbon monoxide levels of less than 5 percent who were comatose. The same facility has admitted patients with blood CO levels of 50 percent who were both ambulatory and could speak.

Duration of Exposure. The longer a person is exposed to carbon monoxide, the higher his COHb level will rise until equilibrium is established with the ambient level of CO. According to the EPA, "Continuous exposure to 30 ppm of carbon monoxide leads to an equilibrium COHb level of 5 percent. About 80 percent of this value occurs in four hours, and the remaining 20 percent over the next eight hours. Continuous exposure to 20 ppm of carbon monoxide leads to a COHb level of 3.7 percent, and exposure to 10 ppm results in a COHb level of 2 percent. Levels as low as 2.5 percent COHb have been shown to aggravate symptoms in angina pectoris patients." (*Introduction to Air Quality—A Reference Manual,* EPA, July 1991.) The reduced flow of oxygen to the heart due to CO exposure causes the deterioration and death of heart tissue, resulting in heart disease. In addition, the results of a study by the North American Congress of Clinical Toxicology found that long-term exposure to low-level CO may contribute to lasting brain damage.

Half-Life. Another pertinent aspect of carbon monoxide is its half-life. Besides breathing it in, we also expel it in the air we exhale. The problem is that we give off CO more slowly than we take it in. This causes retention of CO in the system, which can lead to accumulation of carbon monoxide in the blood until equilibrium is reached with the ambient CO level.

One-half of the carbon monoxide a person absorbs into his bloodstream can take as much as five hours to leave his system. If that person has a carboxyhemoglobin level of 10 percent, it will be reduced to 5 percent after five hours of breathing fresh air. The speed of this process is increased by administering pure oxygen, and it is increased even more if hyperbaric oxygen is administered. In the presence of pure oxygen, the half-life of carbon monoxide is eighty minutes. Given hyperbaric oxygen, the half-life drops to twenty minutes.

The more carbon monoxide that is in the air you breathe, the higher and faster your COHb level will rise. Altitude, too, is a factor. A 3 percent COHb level at a mere 1,400 feet is equivalent to 20 percent COHb at sea level. The faster you breathe in CO, the faster your heart will pump it through your system, and consequently, the faster your COHb level will rise. An increase in activity will increase your respiration rate and heartbeat. Even doing light work in a CO-contaminated atmosphere can greatly increase your uptake of carbon monoxide. Also, the smaller the victim's body mass, the greater and faster the effect of CO on his system. Young children and small pets will be affected sooner and more severely than normal, healthy adults.

Any medical condition that reduces the oxygen available to a person's system will cause him to be more severely and more quickly affected. Those with heart and lung problems, as well as those with anemia, fall into this category. The gas affects different parts of the body differently. Tissues that have the greatest need for oxygen, such as the heart wall muscle and the brain, are the most quickly and most adversely affected. This makes anyone with a diminished oxygen intake more susceptible to CO poisoning. With COHb levels as low as 10 percent, the cardiovascular system of a susceptible person can be stressed and may incur physiological damage. This group of susceptible persons includes those with an existing heart or lung condition, the elderly, small children, and the developing fetus. Anyone with an existing heart or lung condition is already operating on reduced levels of oxygen. Since the additional reduction caused by breathing CO will affect him sooner, such a person may need medical treatment after a CO exposure, even if everyone else in the home is essentially fine. The elderly often have heart and lung problems, as well as other medical conditions, and many of them are nonambulatory. This forces them to spend more time indoors than the general population, meaning that they will get a larger dose of carbon monoxide than those who go out of the home for a number of hours a day. As mentioned above, small children have less body mass than adults, and they also have a faster metabolism, making them more susceptible to the effects of carbon monoxide. Like the elderly, small children often spend long periods of time in the home, and so may be exposed to CO for longer periods than their parents and older siblings will. Still, they may not be able to articulate the symptoms of early poisoning, and so the condition may go unnoticed. At the extreme is the developing fetus, whose blood has a greater affinity for CO than does the blood of its mother. A pregnant mother with CO in her blood delivers less oxygen to the fetus, and the CO in the fetus's blood has a longer half-life than it does in the mother. Its small body mass and rapid metabolism make the fetus particularly susceptible to the damage done by CO. As the mother's carbon monoxide level rises, the level of CO in the fetus rises, but more slowly. If the mother is removed from the exposure and allowed to breathe clean air, her COHb level will start to drop. The COHb level of the fetus, however, will continue to rise, ultimately reaching a level as much as 15 percent higher than the peak reached by the mother. In addition, when its COHb level begins to drop, it will drop more slowly than the mother's. Since the half-life of fetal COHb is three to five times longer than the half-life of maternal COHb, the mother could be just fine after exposure to CO while the COHb level of the fetus remains dangerously high.

SYMPTOMS

A victim of CO poisoning probably won't be aware that he has a problem or that the problem is carbon monoxide. As a first responder, you are in a position to diagnose and initiate early treatment for CO poisoning, but to do so, you must know its symptoms. I remember being taught that such a victim will exhibit a cherry red coloration, but this is not a reliable indicator, since by the account of doctors who treat CO patients, only about 10 percent of victims manifest this change. It is more likely that the patient will be cyanotic due to the deprivation of oxygen. The victim will present other recognizable symptoms long before he appears cyanotic, however, and these are the ones that you must learn to recognize.

Mild Exposure

Mild carbon monoxide poisoning mimics the flu. It causes a headache, usually a frontal headache, as well as nausea, vomiting, and a feeling of fatigue. Small children may suffer intestinal discomfort and diarrhea before adults present any symptoms. Since diarrhea isn't normally considered a symptom of CO poisoning in adults, you might overlook it as a warning sign of CO poisoning in children.

If your child told you that he had a headache and was nauseous; that he had vomited and felt tired, what would you prescribe? You'd probably prescribe bed rest and plenty of fluids, yet if it were CO poisoning and not the flu, you would have given him the worst possible prescription. You would have guaranteed that his condition would become more severe by keeping him in the contaminated atmosphere.

Recently, a family of five in Cleveland went to the hospital complaining of flulike symptoms. The doctor assured them that they weren't suffering from CO poisoning and instead diagnosed them all as having the flu. He sent them home to rest, and three days later, they were all found dead, victims of CO poisoning.

One study found that 23.6 percent of studied patients presenting at a hospital with flulike symptoms actually suffered from low-level CO poisoning. According to a First Alert press release, several recent medical studies show that patients who are treated for CO poisoning often have family members at home with similar flulike symptoms. If someone is brought to the hospital with CO poisoning, it's a real possibility that someone else is at home in worse condition. If a thorough search for additional victims hasn't been made, the result may be another newsworthy tragedy.

It's estimated that as many as one-third of the cases of CO poisoning go undetected, while CO is responsible for 10,000 illnesses a year that are serious enough to require the loss of workdays. It can also mimic food poisoning, psychiatric illness, migraines, stroke, substance abuse, and heart disease, thereby earning its reputation as the great imitator.

If one person in the family has these symptoms, it may be the flu. If the entire family or several families in an apartment have symptoms, consider CO. The flu usually affects one or two family members at a time, then skips around until all of the members have been infected at different times. Carbon monoxide poisoning can affect the entire family at once.

Medium Exposure

A medium exposure to carbon monoxide will cause severe headache, drowsiness, confusion, and a fast heart rate. The fast heart rate is a result of the body trying to compensate for the lack of oxygen that the victim is experiencing. The drowsiness and confusion can prevent a victim from removing himself from the contaminated atmosphere. In one instance, a man and a woman sat down in front of their fireplace after an active day outside in the cold. Unbeknownst to them, carbon monoxide was escaping into the home from the fireplace, and soon they fell asleep. The man woke up and, realizing something was wrong, tried to pick up the phone to call for help but was unable to do so. He tried to get out the door to go for help. The last thing he remembered was falling to the floor.

He awoke outside to find his wife lying next to him. They were both being treated by medical personnel. Luckily, a neighbor had found him passed out on the porch and had called an ambulance. The man didn't remember opening or going through the door.

Carbon monoxide poisoning causes confusion and loss of motor control, so simple tasks become impossible. Tasks such as picking up the phone to call for help or opening a door may not be possible. The victim may slur his speech and speak incoherently, leading you to believe that he is under the influence of drugs or alcohol, rather than in need of medical attention for CO poisoning.

Severe Exposure

A severe exposure will ultimately cause unconsciousness, convulsions, and death. It can occur as a result of a slow buildup of carbon monoxide, or it can occur quickly with no warning after only a few breaths of high concentrations of the gas. The latter has occurred to firefighters who entered smoky, contaminated cellars without the protection of SCBA. In one such incident, a firefighter entered a cellar in which there was a haze of smoke. He didn't wear a mask, and after several breaths of CO-laden air, he fell to the ground. Another firefighter rushed to his side and suffered the same fate. It took mask-equipped members to remove them both to safety.

SYMPTOMS OF CO POISONING BY PERCENT COHb

Exposure	Symptoms	% COHb in Blood
Mild	Slight headache, nausea, vomiting, fatigue.	20–30%
Medium	Severe headache, drowsiness, confusion, fast heart rate.	30–50%
Severe	Unconsciousness, convulsions, death.	50–70%

TREATMENT

As a first responder, you must not only recognize the symptoms of carbon monoxide poisoning, you must also be able to treat its victims promptly and properly.

Fresh Air

The single most important thing that you can do for a victim of carbon monoxide poisoning is to remove him from the contaminated atmosphere. Once in fresh air, he will begin to purge the deadly gas from his blood. His COHb level will begin to drop according to the parameters of its half-life vis-a-vis the ambient environment. As mentioned above, in fresh air, the CO level in his blood will be halved in about five hours.

Oxygen

Many fire departments carry oxygen, and they should treat carbon monoxide victims with it. With 100 percent oxygen and a tight-fitting mask, the half-life of bloodstream CO drops to about eighty minutes. The quicker you can remove CO from the victim's system, the less oxygen deprivation he will have to endure and the less tissue damage he will suffer.

Hyperbaric Oxygen

Consider using hyperbaric oxygenation when a serious CO exposure has resulted in unconsciousness. In the case of pregnancy, hyperbaric oxygenation may be employed, but such treatment should be predicated on consultation with the hyperbaric facility and an obstetrician, as well as the patient, if she is capable. You must be able to recognize the need for hyperbaric oxygen and alert the medical personnel at the receiving hospital to that need. A transfer to a hyperbaric-equipped institution may be required, and the logistics of such a transfer may take time. Early notification of the need can reduce the time required to set up the transfer. Remember that even hyperbaric patients can later suffer a relapse as CO is slowly released from muscle tissue. All such victims must be monitored for some time after treatment.

SUMMARY

Our primary goal is always to save life. In the case of carbon monoxide incidents, we must recognize the symptoms and properly treat its victims. The action we take from the time we discover a victim to the time we release him to hospital personnel may mean the difference between life and death. If we as first responders don't recognize the dangers or don't treat CO-related runs seriously, we may well miss the warning signs and take no action.

STUDY QUESTIONS

1. A hemoglobin molecule has how many binding sites for oxygen?

2. As the body is deprived of oxygen, the body tries to compensate by increasing the _____.

3. Given continuous exposure to carbon monoxide, a person's COHb level will rise until equilibrium is established with _____.

4. A 3 percent COHb level at a mere 1,400 feet is equivalent to _____ percent COHb at sea level.

5. The smaller the victim's body mass, the _____ and _____ the effect of CO on his system.

6. What tissues of the body are the most quickly and most adversely affected by carbon monoxide?

7. Is cherry red coloration of a suspected victim a reliable indicator of CO poisoning?

8. Mild CO poisoning mimics the symptoms of _____.

9. With 100 percent oxygen and a tight-fitting mask, the half-life of bloodstream CO drops to about _____.

10. True or false: Hyperbaric patients can later suffer a relapse as CO is slowly released from muscle tissue.

Chapter Eleven
The Carbon Monoxide Emergency

Carbon monoxide is slightly lighter than air, having a vapor density of 0.96. Besides being colorless, odorless, tasteless, and toxic, it is also *explosive.* This aspect of the gas isn't normally a problem for us when we respond to a sounding CO detector, however, since carbon monoxide's explosive range is between 12.5 and 74 percent, while it is rapidly fatal at only 1.28 percent. Therefore, if anyone in the home is alive when we respond, then the atmosphere won't be explosive. We expect to encounter explosive levels of carbon monoxide at structure fires in confined spaces, but it would be highly unlikely to encounter such a level emanating from a home appliance.

SOURCES IN THE HOME

Carbon monoxide is produced as a result of the incomplete combustion of carbon-based fuels, including wood, natural gas, propane, oil, and kerosene. Complete combustion of these fuels results in the production of CO_2 and water, but if the burning fuel isn't consumed completely, deadly carbon monoxide gas is formed. In reality, fuels never burn completely and, as a result, carbon monoxide is routinely produced in the home. When everything is working properly, this toxic gas is formed in harmless quantities, passing up and out of the appliance's flue. Unfortunately, things don't always work properly.

Fuel-Burning Appliances. All appliances that use carbon-based fuel produce some CO. Solid fuels create the most carbon monoxide, liquid fuels generate less, and gas fuels produce the least. A fireplace, for instance, gives off a certain amount when the fire is roaring in the hearth, with the log being destructively reduced by heat and the resultant gases burning above it. As the log burns down, however, the glowing embers give off more CO than did the burning gases.

One source of CO that you might miss is a gas-operated refrigerator. Look for it in vacation homes, camping vehicles, and anywhere else that you might find people living without electricity. You might also find one in the home of an elderly person who simply never bothered to upgrade to an electric model.

Poorly Maintained Appliances. A properly maintained and adjusted appliance will produce less carbon monoxide than one that is malfunctioning or poorly maintained. An appliance that is clean and has a properly adjusted flame height, as well as an adequate airflow, produces less CO than it would otherwise. Clogged air vents that restrict the flow of oxygen to the flame will cause increased CO production. A dirty or clogged burner or fuel jet, or an improperly adjusted flame, can result in a cooler flame and thus increased production of CO.

Cold Appliances. All fuel-burning appliances produce more CO when they are cold and less when they have properly warmed up. Before an appliance reaches operating temperature, excess CO may spill into the house. A cold pot, for instance, can increase the formation of CO when it is placed on the burner of a gas stove. This is normal and not a health problem. It shouldn't increase the overall CO concentration in the home, but the level of CO near the appliance will rise until the pot warms up. The excess CO produced as a result of a cold pot should harmlessly dissipate. Once the pot warms up and the flame reaches operating temperature, less CO will be produced. Anything that lowers the temperature of the appliance's flame will cause an increase in the production of carbon monoxide. For example, rust flakes on a gas hot water heater's burner will lower the temperature of the gas flame, resulting in the production of excess carbon monoxide.

Inadequate Ventilation. Space heaters, stoves, ovens, and barbecues all generate carbon monoxide and, if not properly vented, will introduce it into the home. Additionally, using an unvented fuel-burning appliance in a sealed home can reduce the oxygen level, and the reduced O_2 supply will cause an increase in the production of CO. Some space heaters use an oxygen depletion sensor to prevent the formation of CO. If the oxygen is sufficiently reduced, the appliance shuts off. Insufficient oxygen, however, isn't the only reason that CO is produced.

A properly functioning gas stove can give off as little as 35 ppm or as much as 400 ppm of CO before reaching operating temperature. A gas oven can give off as much as 800 ppm before coming up to temperature. A stove, however, isn't expected to operate continually, so the CO it gives off shouldn't significantly affect the home's air quality. If, however, the stove is used to heat the home, as it might be if a winter storm knocks out the home's heating unit, then it could alter the ambient CO level, creating a hazardous condition. In this case, the stove would be operating for a much longer time than it was designed for, and the oven door would be left open, changing the designed airflow to the flame. A dirty oven, too, can reduce the operating temperature of the flame, as can lining the bottom of the oven with aluminum foil. Even an improperly adjusted pilot light has been known to raise the ambient CO level to as high as 60 ppm overnight in a sealed home. All unvented fuel-burning appliances spill some CO into the home and, if improperly adjusted, maintained, or operated, will increase the amount of CO they produce.

This badly perforated flue pipe was leaking deadly carbon monoxide into the home. (Photo by Pat Coughlin, IAFC.)

Flue Length and Condition. A properly constructed and maintained flue will vent carbon monoxide to the outdoors. If, however, the flue is blocked by soot, broken masonry, a bird's nest, or other debris, or if it is too long or made up of undersized piping, carbon monoxide and other combustion gases will spill into the home and not be drawn out as intended. A flue might be blown apart by an oil burner puffback or inadvertently knocked out of place by a workman, thereby allowing carbon monoxide to escape into the home. If the flue has suffered rust damage, it might be perforated and leak gas into the home as a result.

Downdraft. This phenomenon is often referred to as backdraft in articles written outside of the fire service. Since firefighters associate the term backdraft with a smoke explosion, I will use the term downdraft in this instance.

Ideally, a home should be operating at neutral pressure. There should be no pressure difference between the interior and exterior of the home, but if the home is tightly sealed, and if air is pulled or pushed out of it, a condition of negative pressure can be created inside. If this negative pressure is strong enough, it can result in flue gases being sucked back down and into the home.

A strong wind blowing across an unused chimney can cause negative pressure in the home. The passage of wind across the opening creates the venturi

Wind blowing across a chimney top draws air up and out of the house, if the damper is open. The resultant negative pressure inside pulls gases back down the flue of the heating unit and into the home.

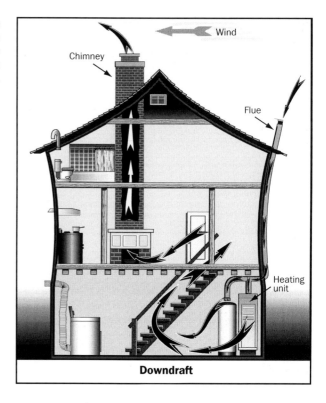

Downdraft

Various electric fans vent exhaust to the exterior of a home. This can create a negative pressure inside, drawing gases back down the flue of the heating unit.

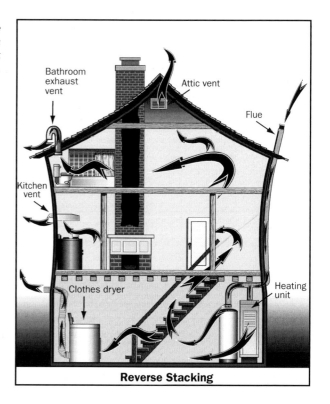

Reverse Stacking

effect, which can suck air up and out. Air can also be drawn out of a house if there are windows open on the leeward side. If the negative pressure created is stronger than the natural draft of appliances such as your furnace or hot water heater, carbon monoxide will spill back into the home instead of being drawn up and out the flue.

Reverse stacking is a mechanically induced downdraft. In the modern, tightly sealed home, there is a constant struggle between appliances for air. Your furnace needs air for combustion, as does your water heater. At the same time, your clothes dryer is using air for combustion and pushing air out of the home through its vent. Your bathroom and kitchen fans, too, are pushing air out of the home, and if you have an attic fan, then it is expunging a potentially enormous quantity of air.

It can be said that air wars are constantly being waged in the home, and most of the time, fans win these wars. They push more air out of the home and with more force than do fuel-burning appliances. The result can be reverse stacking, or the sucking of flue gases back into the home. If the water heater and furnace are enclosed in a small room, reverse stacking is possible unless enough makeup air is allowed into the room. Typically, the furnace will draw flue gases from the water heater into the room rather than allowing them to go up the flue. This is especially likely if they are ducted into the same flue.

Downdrafting and reverse stacking can be prevented by providing makeup air for the appliances in use. The simple act of opening a window can prevent the flue gases from entering the home by providing makeup air and equalizing the pressure inside and out. On the other hand, closing a window or door can sometimes cause downdrafting or reverse stacking.

If, by the time the fire department responds, the weather conditions change and the wind is no longer blowing as strong as it was when the downdraft occurred, you may not be able to determine conclusively why carbon monoxide entered the home. If the fans that were running are no longer running, you may no longer have an appliance spilling back CO.

Flue Temperature. As the temperature of the flue gas is reduced, the draft is reduced. Thus, the appliance becomes more susceptible to downdrafting and reverse stacking. Smaller, more efficient heating units are designed to prevent the escape of excess heat up the flue and instead to use it to heat today's homes more efficiently. The heat going up the flue, however, is partially responsible for the draft that sucks waste gases out of the home. At some point, the high-efficiency furnace can become so efficient in removing heat from flue gases that it will need a fan to create the draft. The upgrading of an old furnace might remove 65 percent of the heat going up the flue and the draft will be fine, whereas at 80 percent reduction, the same furnace might not have enough draft to prevent spillage. In this way, a furnace that didn't spill carbon monoxide might be improved enough to become a CO problem. To avoid this, fans are used to induce a draft in the flue. The danger lies in the heating unit that has had its efficiency increased but that hasn't had an exhaust fan added.

Another related problem is that, as a result of the cooler gas temperature, the acidic flue gases condense and cause the flue to deteriorate. The acidic condensate can cause perforations in a metal flue, which will leak CO into the home. High-efficiency furnaces require the use of stainless steel flue pipe to reduce the effect of the acidic condensate on the flue pipe.

This condensation has caused another problem in arctic climates. When these cooler flue gases are subjected to sustained subzero temperatures, the condensed moisture can find its way into the lining of the flue, causing it to collapse inward as the moisture freezes and expands. This ice continues to build up throughout the winter until the flue has been partially or completely blocked. In one instance in Yellowknife, the capital of the Northwest Territories, a mother and her six children were sickened by carbon monoxide that had entered their home as a result of a collapsed chimney liner. The instigator was condensation that had seeped into the chimney liner, where it had frozen, expanded, and blocked the flue. In extreme climates, the flue can, in time, freeze solid like the trunk of a tree, and it will have to be cut out and replaced.

Vehicles. Automobiles are a potentially deadly source of carbon monoxide in the home. Cars left running in attached garages have been responsible for many CO-related deaths. Autos give off extremely high levels of carbon monoxide while they are warming up, and warming up a car in an attached garage can cause CO detectors in the home to go off even if the exterior garage door has been left open. If there is a door or other opening between the house and garage, CO can be drawn into the living portion of the home by negative pressure. A car that hasn't been warmed up can initially give off 30,000 ppm or more of carbon monoxide. After warming up the car in a garage with its exterior door open, as much as 200 ppm of CO was recorded on the dwelling side of a door common to both the garage and the home.

Tightly Sealed Homes. The installation of new windows or even weather stripping can change the airflow in your home and create a carbon monoxide problem where none existed before. In one case, a ceremonial Indian hut called a hogan, traditionally made of logs and mud with a smoke hole in the roof, was turned into a CO death trap. It was made of modern building materials rather than the traditional ones, and it was attached to the home of a Navajo woman. In the hut, a fire was lit as part of a religious ceremony, but instead of the smoke and other pollutants seeping out of the openings usually found in such a structure, they were held in while fresh air was held out. Three people died as a result of CO poisoning.

As we seal our homes against winter's cold and summer's heat, we seal in pollutants and keep out fresh air. The normal airflow or breathing of a home is stopped. Thus, any CO produced in the home can't escape and therefore builds up. Although weatherproofing homes saves money on energy costs, it may create a problem that could cost the occupant his health or even his life.

Charcoal. Burning charcoal indoors will contribute potentially deadly levels of CO into the home. For this reason, charcoal manufacturers must include on their packaging a warning against burning it indoors. Many years ago, I felt the effects of CO poisoning when, on a November camping trip in the mountains, I brought a small hibachi into a tent for warmth. The burning charcoal gave off fumes that made me sick. Luckily, I realized what the problem was in time. I removed the hibachi, and the symptoms disappeared.

A charcoal grill next to an open window or set up in an attached garage on a rainy day can be enough to trigger a CO alarm in the home. Cooking or heating with charcoal inside a home or tent can be deadly.

Exterior Sources. Externally produced carbon monoxide can cause a deadly buildup of CO in your home. If your neighbor has a CO problem in his apartment, the gas can seep into your apartment. If your home is attached to one that has a CO problem, you too may have the problem. A running car or truck parked near an intake fan or air conditioner can pour CO into your house. In one instance, many people in an airline terminal were made sick. An investigation revealed that a luggage vehicle had been parked near an air intake and that exhaust fumes had been sucked into the terminal. Had the vehicle been moved before the investigation took place, the cause might never have been discovered.

DISPATCHING THE ALARM

Responding to CO alarms isn't as exciting as responding to and operating at fires, but it is something that you will probably do more often than fight fires. In 1993, approximately 18,000 CO detectors were sold. That number rose to 373,000 in 1994 and to more than five million in 1995. In 1991, Kingston, New York, passed legislation mandating the use of CO detectors in dwellings with more than four apartments. Chicago, responding to the death of a family of ten, also passed a mandatory CO detector law. Both New York City and New Jersey have considered such laws.

Today, approximately six percent of the homes in the United States are equipped with CO detectors, and the estimated potential market is 200 million homes. You can bet that the number of home CO detectors will continue to rise as cities pass mandatory detector regulations, as manufacturers pursue aggressive advertising campaigns, and as CO deaths continue to be reported by the media.

Earlier I mentioned that fire departments need to learn how to respond to CO alarms. How is such a response different from fire-related responses? When we respond to fires, our hearts race as the adrenaline surges, charging us up to do battle with the devil. This is appropriate for a fire response but inappropriate for a CO alarm. For such a response, we need not be gearing ourselves to exert a superhuman effort. Instead, we should be thinking like detectives and medical personnel. Home CO alarms are designed to sound before the levels of the gas become deadly, so it is unlikely that your response will require immediate emergency action. Not impossible, but unlikely. The majority of such responses will be the result of an alarm responding to a nonlethal level of gas, and it will require investigative skills to determine the source and medical skills to diagnose and treat the occupants.

Before you respond to a CO call, decisions have already been made that may affect the outcome of the response. These decisions have been made by your dispatchers when they receive the call for help. How many units do they dispatch? Do they dispatch an ambulance? Do they exercise discretion in deciding what units to dispatch? *Should* they exercise discretion?

To answer these questions, first we must decide what is an appropriate response to a CO alarm. How many firefighters will be needed? What apparatus is needed? How you answer this question will determine the number of CO calls your units will respond to and how many units you'll have available for other responses. Since the dispatcher is the first contact that your department has with the caller, he can play an integral role in a CO response. With training, your dispatcher can decide whether to send a full response plus an ambulance or a single unit equipped with a CO detector.

Dispatch Considerations

Is This an Emergency Response? Although a dispatcher can't positively determine this, he can, by asking a few questions, get a pretty good idea as to whether anyone in the home is in danger. The dispatcher should ask the caller whether he or anyone else in the home is feeling the symptoms of CO poisoning. If no symptoms are evident, then it isn't an emergency, and one unit may be able to handle the call. The dispatcher should listen carefully to the caller's responses. If the caller is incoherent, disoriented, or slurring his speech, he may be suffering from CO poisoning. He may also be under the influence of alcohol or drugs. Since the dispatcher can't tell which is the case, he must consider the situation to be an emergency and dispatch the appropriate units. In any event, when the dispatcher is in doubt, he must consider the call an emergency and dispatch units accordingly. If the dispatcher determines that no one is feeling sick, and if the caller is alert and responsive, then the call need not be considered an emergency.

What Should the Caller Be Instructed to Do? If anyone in the residence is feeling symptoms, the dispatcher should instruct the caller to do a head count and then have everyone leave the building and not return until the fire department has deemed it safe to do so. If no one is feeling symptoms, the dispatcher can instruct the caller to open windows, ventilate the home, and shut down all sources of CO. It isn't necessary for the occupant to exit the building if no one is feeling symptoms.

What Units Must Be Dispatched to a CO Alarm? A single unit can handle the response if no one is sick and the building is relatively small. However, if people are in need of medical care or an investigation and possible evacuation are involved, additional units must be dispatched. For any indication of symptoms, the dispatcher should consider sending one or more ambulances. The first-arriving fire unit on the scene will determine exactly how many ambulances are needed. It is the on-scene fire officer who should make the final determination as to what constitutes the appropriate response.

I have spoken to several departments on the subject of allowing the dispatcher to decide what units to send to a CO alarm. Some departments already do it this way successfully. Others don't feel that their dispatchers should be given a choice as to how many units to send. You must consider how well trained your dispatchers are and how well they know your department's procedures, capabilities, and needs before allowing them to make such a response decision. In any case, it is essential that the dispatcher be kept apprised of the status of each unit's CO meter so that a meter-equipped unit can be sent. When a unit's CO meter is out of service, the dispatcher should be notified. If that unit is dispatched to a CO alarm, the officer should notify the dispatcher of their status and request that a unit with a viable meter also respond.

How Well Trained Are Your Firefighters? Is a chief needed on these calls, or are the company officers and firefighters trained to handle them? A well-trained unit doesn't need a chief officer to determine what action should be taken in response to a home CO alarm. If the incident escalates to become a true emergency, you may have to call the chief, especially if you find many victims, if the building is large, or if the search will be extensive. Find out what departments of similar size in your area are dispatching to these alarms. What works well for them might work well for you, too.

FIRE DEPARTMENT RESPONSIBILITY

Our task at these responses is to locate the source of the carbon monoxide, eliminate it, and leave the home safe for its occupants. Typically, the amount of CO that sets off an alarm isn't lethal. Why then, you might ask, should you bother trying to locate and eliminate a nonlethal source? The answer is that medical evidence indicates that long-term exposure to low-level CO can have lasting harmful effects on the body. It is believed that such long-term, low-level exposure can actually cause heart problems in a healthy person. Extended exposure to low levels of CO is also suspected of causing lasting brain injury, according to a recent study by the North American Congress of Clinical Toxicology.

Another reason is that a seemingly nonlethal, low-level source may later become lethal. For example, you might respond to any alarm caused by a furnace that's leaking just enough CO into the home to bring the ambient level up to, say, 10 ppm for a prolonged period of time. This isn't life-threatening, but if the source isn't found and eliminated, and if the temperature drops, the increased activity of the furnace might contribute more CO, possibly in deadly amounts. As the temperature drops, the furnace will operate continuously, burning more fuel and possibly spilling more CO into the home. If, when you investigate, you find only low levels of CO, tell the occupant that it is a nuisance alarm, and then depart without searching for the source, you may be leaving him in danger. What if, later, the source becomes deadly? What will happen if the occupant dies as a result of CO poisoning? Remember, you responded and declared the situation safe. Did you perform a detailed examination to locate the source? Did you document it? Can you defend your actions in court or explain them to surviving family members?

Tennis star Vitis Gerulitis died as a result of CO poisoning. His COHb level was 70 percent when he was found dead by a maid in a guest house in the Hamptons on Long Island. The emergency personnel who responded took readings but found no carbon monoxide in the building. The source of the gas had been a pool heater that operated intermittently, so there wasn't a constant presence of it in the home. It was, however, enough to kill the athlete.

What if the fire department had been called to the guest house earlier in the evening in response to a sounding CO alarm? What if personnel had responded, found no threat, and told the occupants that the alarm was defective? What if Gerulitis had then disabled the alarm, based on their statement? Do you think that the fire department would have been named in the 67 million-dollar lawsuit that ensued? Now think about your last CO alarm response. Could you defend your investigation and its results against such a suit?

DETECTION DEVICES

You must have the appropriate detection device and be trained in its use. Responding without a gas meter capable of detecting carbon monoxide is a waste of time. The only indication you might have would be medical symptoms. The only actions open to you will be either to leave a potentially deadly source of CO still functioning in the home or to shut down all appliances, leaving the homeowner without heat, hot water, or the use of his stove.

If you don't have a CO meter, you will have to request that another agency or unit respond. The appropriate device to use is a meter that can detect CO and display a continuous digital readout of the concentration. It can be a multigas meter or a single-gas meter, depending on your department's needs and resources.

This single-gas CO meter reads from 0 ppm to 999 ppm.

Single-Gas Meter. Single-gas meters are available that don't even need to be turned on. They are left constantly on and have a battery that is replaced monthly. These simple meters perform well in detecting and measuring carbon monoxide. In addition, they are easy to read and operate.

Multigas Meter. Multigas meters are more expensive than single-gas meters, but their ability to detect other gases makes them eminently more useful. The multigas meters that I tested were capable of displaying several types of readouts and could perform various functions. I found that, for a CO response, the exotic functions weren't needed.

This multigas meter reads CO plus three other gases of your choice. Because of its flexibility, you can set it up to suit your needs at any given incident.

If your funding permits, a datalogger/meter can be useful in tracking down hard-to-locate sources. This type of meter can be left in a home. After several days of monitoring the ambient air, it will produce readings that can be downloaded into a computer and used to print a graph of the gas levels and times of occurrence in the home. It is an excellent tool for finding an elusive source of carbon monoxide.

Purchasing a number of multigas meters might be cost-prohibitive. The cost of one, which is in the two-thousand-dollar range, could buy several single-gas meters, which are available for as little as two hundred dollars each. If you plan to purchase only a few meters, the versatility of a multigas meter might be the better choice for you. If you plan to purchase many meters, the majority of your companies could be equipped with single-gas meters. You could then equip specific units, such as rescue, haz mat, and chiefs, with multigas meters, making them available for confined-space rescue and other gas-monitoring needs, as well as for CO investigations. Checking with nearby departments and finding out which meters suit their needs will help you in making your own allocation decisions.

Other Features

Ease of Operation. The simpler to operate, the better. A complex meter may be able to perform all sorts of exotic functions, but it must be operable by your firefighters. Will you be able to train all of them in the workings of a complex gadget, and will competent firefighters be on duty when the company is called to a sounding CO alarm? I prefer the simplest meter for distribution to rank-and-file firefighters.

Instructions. Your firefighters will have to be trained to use the meter. Your manufacturer should provide this training, but will they send a representative to do so? After you purchase the meters, will their representative be accessible to you, and will he promptly answer your questions? Is the documentation given to you by the manufacturer adequate and easy to understand? Will they include a video with the meter or just a confusing manual? How well your members are trained in the use of the meter will determine how effective they'll be at a CO incident.

Digital Display. The digital display should be easy to read. I tested a number of meters, both single- and multigas, and found some displays to be easier to read than others. Look for one that is illuminated for use in dark areas and yet can easily be read in daylight as well. The meters will have a programmable alarm that you can preset to warn you of the presence of a dangerous CO level. Some meters give instantaneous CO readings, while others display increasing levels until a peak is displayed. Both types are satisfactory for investigations, but the instantaneous-display version is preferable.

Meter Range. In order to detect low-level sources, it is important that the meter be capable of sensing gas levels from 0 ppm to at least 200 ppm, and preferably 999 ppm. The ability to read high levels of gas will better enable you to assess the hazard and locate the source. The meter should also indicate if the level is above 999 ppm.

Calibration. These meters need to be periodically calibrated and serviced. Can your department perform this function, or will the meter have to be sent back to the manufacturer? If the meter has to be sent back, will they supply you with a loaner, or will you be without one for the duration? How long will the service take? How often will the service be needed? You want a meter that will give maximum performance with minimum service. I have tested meters that are easily calibrated in the field. They use bottled gases to test the unit's calibration and are easy to adjust to specifications.

Durability. The meter should stand up to abuse. Firefighters aren't known for handling equipment gently. The meter will bump around in the apparatus, it will get wet, and it will fall. Do these misfortunes mean frequent, lengthy repairs, or will the meter keep working? Does it come with a protective case?

Intrinsic Safety. No meter you use should pose an explosion hazard in a combustible atmosphere.

RF Protection. The meter should be protected against interference from radio transmissions.

TRAINING

Once you have the proper tools, you then need the proper training. You need to recognize that carbon monoxide is a serious problem and that sounding home CO alarms are usually reacting to the presence of the gas. It is important that you instill the correct attitude in your firefighters regarding CO investigations. They mustn't view them as nuisances. If they don't treat such calls as the potentially life-threatening incidents that they are, then they are setting themselves up for a tragedy that could have been prevented.

You must recognize the symptoms of carbon monoxide poisoning and know how to treat its victims. You need to know the sources of CO in the home, how to locate them, and how to eliminate them.

Miners once took canaries and mice with them as they descended into the coal mines. These small animals were used as CO detectors. The animals were affected before the miners themselves, so when the animal collapsed, the miners knew it was time to get out. Fortunately, we are more technically advanced today. We have access to sophisticated home detection devices. Armed with such devices and trained in their use, a homeowner can be warned of a potentially dangerous buildup of the gas. Firefighters armed with the proper gas detection equipment, training, and attitude can locate CO sources and eliminate them.

In December 1995, UPI reported that a Cleveland family of five died from carbon monoxide poisoning. They had gone to the hospital, complaining of the flu. One family member asked whether they could be suffering from CO poisoning, and a doctor assured them that they were not. Several days later, they were all found dead in their home.

That couldn't happen now, you might think. Carbon monoxide poisoning has received so much publicity that doctors wouldn't make that mistake today. Unfortunately, such errors are still being made.

Carbon monoxide is a serious problem, and depending on whose numbers you consult, accidental poisoning is responsible for between 250 and 1,500 deaths a year. As many as ten thousand people are made ill by it each year. It is in every home that uses fossil fuel. One Canadian study done by *Progressive Builder Magazine* determined that one in ten homes using natural gas had a carbon monoxide problem over the ten-year period of the study.

The current CO home detection technology is fairly accurate. When a detector sounds, it's usually responding to the presence of carbon monoxide. It may be reacting to low levels of the gas and the situation may not be dangerous, or it may be responding to life-threatening concentrations. To find out which requires an investigation. Remember, a low-level, nonlife-threatening source can turn deadly at a later time, or the source may be an intermittent one.

Your goal should be to locate the source, shut it down, and ventilate the area, rendering the home safe for its occupants while giving immediate treatment to any victims that you find. When you leave the scene, you must leave it safe. Once the source has been shut down and you're sure there aren't any others, you should take the time to explain to the occupant what you've found and done, as well as what he must do to rectify the problem, including calling a repair service or the utility company. Take readings again after ventilating and before you leave to ensure that you haven't missed another source.

RESPONSE PRECAUTIONS

I read an account of a carbon monoxide incident in which an ambulance crew responded to a home whose occupants felt sick. Some time passed, and the ambulance crew didn't report in to their dispatcher. Concerned, the dispatcher sent another crew to the location. The second crew found the occupants as well as the first ambulance crew unconscious in the home, all victims of CO poisoning. Since most ambulance crews carry neither gas-detecting equipment nor SCBA, they couldn't test the atmosphere in the building, nor did they have masks to wear. They may not have suspected CO poisoning, but if they had, the appropriate precaution would have been to call the fire department. Ours is the logical agency to handle these types of responses. Firefighters have the appropriate training and equipment, including tools for forcible entry, if needed. Could the above incident have involved a fire company instead of an ambulance crew? Yes, of course, if they didn't have the appropriate equipment and training or did not take the right precautions.

What level of carbon monoxide is safe for you to breathe, and at what level should you don your mask? There is no pat answer to this question. A number of governmental agencies declare various levels as being safe. Fire departments all over the country have adopted SOPs that indicate mask usage at varying levels of CO exposure.

The EPA is responsible for monitoring outdoor air quality. Its standards allow for a twenty-four-hour exposure to a carbon monoxide level of 9 ppm. The CPSC, responsible for ensuring that the appliances used in the home are safe, allows an eight-hour exposure to 25 ppm. OSHA allows a 50-ppm exposure to CO for eight hours.

Why is there such a variance in the acceptable levels of CO? OSHA is concerned with the workplace. This agency expects that a worker will spend only eight hours on the job, exposed to the 50 ppm. When the worker leaves his job, he will no longer be exposed, and the CO will purge from his system. The EPA, on the other hand, is concerned about extended exposure to ambient CO. Most people spend more time at home than they do at work. Unlike the worker, a child, the elderly, or anyone susceptible to CO cannot just leave his home and allow the gas to purge from his system. To compensate for this prolonged exposure, the EPA sets lower allowable limits than OSHA does.

ACCEPTABLE LEVELS OF CO EXPOSURE, BY AGENCY

Agency	ppm Allowed	Time
OSHA	50 ppm	8 hrs.
EPA	9 ppm	24 hrs.
CPSC	25 ppm	8 hrs.

Fire departments around the country have adopted different mandatory parameters for SCBA use. The Chicago Fire Department mandates SCBA at 35 ppm, while the Fire Department of New York sets 9 ppm as the minimum. Other departments mandate SCBA at all CO investigations, while still others mandate its use at 100 ppm. In deciding its SOPs, your department should consider several factors. First, is the number your department has chosen realistic, and will firefighters comply with it in the field? If it is unrealistically low, the minimum may be ignored. Also, will that number protect you against multiple exposures? A firefighter may respond to several such alarms a night, as well as to fires. Will the cumulative effect of the CO he has been exposed to be detrimental to him? Find out what nearby departments have decided and how it has worked for them. Checking the protocols of surrounding departments will give you insight into your own, and it will also point out any problems that they are having.

Firefighters should conduct their own CO investigations in teams. While conducting a CO investigation, the members should wear their SCBA in a stand-by position and don the face piece should the readings reach the specified level. In addition, it's prudent to have an additional team of mask-equipped firefighters standing by to assist the first in case they run into trouble.

Finally, always be aware that the effects on individuals can vary. For example, at a fatal cellar fire, two firefighters died of carbon monoxide poisoning. One had a COHb level of 31 percent and the other a level of 63 percent. The two firefighters were killed by drastically different amounts of the gas. The following table is offered as a reference. The Percent-in-Air column has been included to put the ppm figure into perspective. Remember, relatively small amounts of carbon monoxide can be deadly.

EFFECTS OF CO EXPOSURE BY PPM

ppm	Time	% in Air	Symptoms
200	120 min.	.02	Flulike
800	45 min.	.08	Flulike
800	180 min.	.08	Death
1,600	60 min.	.16	Death
3,200	10 min.	.32	Flulike
3,200	30 min.	.32	Death
12,800	1–3 min.	1.28	Death

I once responded to a call for a fire in an electrical manhole. It turned out to be much more than that. Before the night was over, I had evacuated several private homes for dangerously high CO readings and discovered the gas seeping into an apartment building a block away as well as into a hospital two blocks away. The CO had entered these occupancies via underground electric cables and ducts. Had I not expanded my search, I would have missed both of these buildings.

In November 1986, an executive who missed a morning appointment was found dead and his wife comatose in their motel room, both victims of CO poisoning. In another room in the same hotel, unknown to hotel and emergency personnel, another couple was suffering from CO poisoning. They slept past their checkout time and finally were able to call for help at eight p.m. They were removed to a hospital, and a check was made of the other rooms in the hotel. Another victim was found, and *then* the hotel was evacuated. The source was a faulty pool heater leaking CO into the ventilation system. This incident occurred before carbon monoxide became a regular fixture in the news. Firefighters investigating such an incident today would certainly, I hope, expand their search after finding the first set of victims. If not, they might need the services of a good lawyer.

RESPONSE PROTOCOLS

Determining the Scope of the Problem

Interview the Occupant. Start your investigation with an interview. You want to know whether anyone is feeling symptoms of CO poisoning. Be sure to check all of the occupants of the residence, and that includes the colicky baby who just got to sleep, as well as the grandmother sleeping in the basement or attic apartment. If symptoms are evident or suspected, take the appropriate action to remove the occupants and treat them with oxygen. If no one is exhibiting symptoms, or as you are treating the victims, you can begin your survey to locate the source of the problem. An acquaintance of mine called the fire department because the home CO detector he'd just purchased was sounding an alarm. He'd been feeling sick for the past several weeks whenever he was home, but when he went outdoors he felt better. Suspecting CO poisoning, he'd purchased the detector. This classic symptom, feeling sick at home and better away, should be discovered by firefighters during the initial interview with the occupant.

Once the symptoms have been addressed, you should continue your interview to determine what has gone on in the home for the past six to ten hours. Ask the occupant what appliances have been used recently, but don't take his word that an appliance hasn't been used. Feeling a space heater or stove may help determine the truth.

If everyone in a family of six has taken a long shower, consider the water heater to be a potential source of carbon monoxide. If it is Thanksgiving and the turkey has been roasting in the gas oven since early in the morning, suspect the oven. In fact, you can expect an increase in CO alarms during the holidays because of the accompanying prolonged use of gas ranges.

Take Readings. Take readings at the door. As you are interviewing the occupant at the door, extend your arm into the home and take a reading with your CO meter. If the levels are safe, continue into the home, taking readings as you proceed further inward. Take readings at various levels in the home. The gas, produced by combustion, will initially rise, but once it cools off, it will diffuse evenly throughout the interior.

Check Appliances. Once you have taken a general survey of the ambient air, and if no source presents itself, you must extend your search. Check each fuel-burning appliance. If you find high readings near one, consider it to be a possible source. Remember, though, it isn't uncommon for an appliance to give off excess carbon monoxide as it warms up. This will increase the levels near the appliance but not necessarily throughout the house. Unless a stove, for example, is being used to roast a large holiday turkey, it will be shut down long before the other rooms show a serious increase in carbon monoxide.

Use a Checklist. It is a good idea to use a checklist when conducting a carbon monoxide investigation. The checklist should list common home appliances and allow for multiple readings to be entered alongside each appliance. Using a checklist will ensure that the investigators don't overlook a possible source, and it can serve as verification that a thorough investigation was performed. It can be incorporated into a form to be left with the occupant. On the form, you can list your findings, what action you took, and what the occupant should do. It's possible, however, that this notice of findings could be used against your department in a court proceeding if it reflects sloppy work.

Expand Your Search. If you find no source of carbon monoxide in the occupancy, you may have to expand your investigation even further. In the Tennessee motel incident mentioned previously, there was no source of carbon monoxide in the motel room. The source was the pool heater. Always be aware that CO can seep in from adjoining apartments, or even adjoining buildings. If such is the case, the CO detector will be sounding in an occupancy in which there is no direct source. I discovered a home that ducted its flue gases into an adjoining, unattached building's chimney. A downdraft or cracked flue liner in the second building could result in a CO problem even if the building's heating system were not functioning. Would your investigation discover such a source? Consider also that the gas might be entering the occupancy through any of its four walls, as well as from above or below. There are six exposures to consider as you expand your investigation. In one instance, carbon monoxide was found to be leaking into the attic from a leaky flue pipe. The CO would build up in the attic, then seep down into the home through a closed pull-down staircase opening, triggering the alarm. Remember that you may even have to take readings in the attic when investigating a CO source in the living space.

When No Gas Is Present. You may not detect any CO. You may respond to a home where an alarm has sounded, but by the time you arrive, there is no

The Carbon Monoxide Emergency • 201

Carbon monoxide can come from sources you might not expect. The flue of the house on the left runs into the chimney of the house on the right. A cracked flue could leak CO into the living space of the latter even if its heating unit is turned off.

longer CO in the residence. Consider the possibility of an intermittent source, or that the occupant might have ventilated the area prior to your arrival.

Get the Big Picture. If you don't find a source in the building, step back and get a look at the big picture. While investigating a CO condition in one private home, we had low-level readings but could find no source. I stepped outside and looked at the detached private dwelling. Locating the chimney, I directed the firefighters to test for CO along the interior-facing sides of the chimney, which were flanked by two closets on the second floor. In one of the closets, we found elevated levels of CO. Apparently, the flue was deteriorated at that location, and as the furnace heated the home, the flue gases had seeped into the closet. As the gases built up in the closet, they spilled out into the room, causing the detector to activate. If the weather had turned colder, the furnace would have pumped in even more flue gases, raising the ambient level of CO throughout the home, possibly to deadly levels. The IAFC, in a carbon monoxide update, stated that CO levels in closets and cabinets may remain high even after the home has been ventilated. It's likely that this will occur after a high level of CO has built up in the home. Taking readings at these locations may prove that there had been a buildup of CO in the home before it was ventilated. Such a discovery should be considered proof that CO was present in the home, inducing firefighters to extend their search for its source.

Even if carbon monoxide is present in the home, you may not be able to locate the source. A number of variables can make a source difficult to locate. One such hard-to-find source is downdraft.

Set Up a Worst-Case Scenario. Often you'll find no carbon monoxide in the home when you respond. Possibly the occupant has already vented before your arrival, or the source is no longer active or it is intermittent, and the natural airflow in the home has removed the poisonous gas. It's easy at this point to tell the resident that he has a defective alarm. The problem is, without further investigation, you can't be sure that the alarm is defective, nor can you be sure the home is safe.

If, after a thorough search of the house for carbon monoxide and defective appliances, you find nothing, there is more that you can do. You can set up a worst-case scenario:

1. Turn on all fuel-burning appliances. Light the oven and stove burners, and if a space heater is present, turn that on, too. Try to re-create the conditions that might have been present in the home before your arrival.

2. Close all of the windows and doors to seal the house as tightly as possible.

3. Set the thermostat high enough to activate the heating unit.

4. Run the hot water until the water heater turns on.

5. Start the gas clothes dryer. This appliance not only creates CO, it also vents air outside the home, which may help create negative pressure inside and may induce reverse stacking.

6. Turn on the house's various fans. This includes the kitchen and bathroom fan, as well as the electric clothes dryer and the attic fan, if present. Again, these may contribute to negative pressure and reverse stacking.

Once you have done all this, take readings throughout the house, and then take readings again after all of the appliances have warmed up. By re-creating the carbon monoxide problem in the house, if there is one, you may identify a source that you couldn't previously locate.

If Necessary, Use a Datalogger. For those instances when you cannot find the source, especially if you are repeatedly called to the home, a datalogger is indicated. A datalogger is a carbon monoxide meter that can record CO activity in a home over time and later download the information into a computer. The resultant graph can show when CO levels rise, as well as how much of the gas is present at various times. This, coupled with an analysis of the activity in the home, plus the weather conditions, may help point investigators toward the source.

Call for Help. If you're unable to locate the problem and don't have a datalogger, you may have to call for help. Your local utility company or oil burner repair company may have a datalogger available, or there may be other government agencies that can assist by monitoring the CO level in the home. You should make arrangements in advance to obtain the assistance of such agencies, and you should have the names and numbers of those agencies and businesses that can help you locate the problem. Some gas utility companies will leave a datalogger in the customer's home to monitor the CO levels over time. If the meter isn't returned, they simply add the cost of it to the person's gas bill. For we in the fire service, it wouldn't be as easy to recoup the cost of a lost, damaged, or stolen datalogger from an occupant, and so it might not be practical for us to provide this service.

Signs of Carbon Monoxide

Once you are in the home and conducting your investigation, be alert to any outward signs of the gas. Bear in mind that the only conclusive way to detect it and locate its source is with a CO meter.

Odors. Carbon monoxide is odorless, but the presence of stuffy, stale, or smelly air may indicate a CO problem. These are indicative of a house that is tightly sealed and with little, if any, air exchange. An aldehyde odor may be discernible if the source is a gas furnace. If the source is an oil burner, you'll probably notice the telltale oily smell.

Moisture. Moisture on windows is an indication of high humidity. Water is one of the products of the combustion of fossil fuel. If a gas burner is leaking flue gases into the home, you might notice moisture collecting on the windows. Of course, the moisture might also be no more than the result of cooking vapors meeting a cold window pane.

Flue Gases. Soot around the fireplace, furnace air intake, or draft hood of a gas-fired furnace or water heater indicates that gases are backing down the flue and spilling into the home.

Gas Flame Color. A properly adjusted gas flame will burn blue once the appliance reaches its operating temperature. A gas flame that's yellow or orange is one that isn't properly adjusted and is giving off excess CO. Yellow or orange coloration may be normal as the unit warms up, but once it reaches operating temperature, the flame should be blue. Gas fireplaces are typically unvented, and they can give off high amounts of CO. The gas flame, normally blue, is engineered to be yellow or orange to create the cozy atmosphere of a real wood flame. This yellow flame is created by limiting the amount of air available for combustion, but it brings with it increased CO production. Some gas fireplaces and other appliances have an oxygen sensor that will shut down the gas flame if the oxygen in the room is depleted. The problem is that CO can be produced even if the O_2 isn't depleted. Some newer gas fireplace logs use a chemically coated rod to create a flame of the desired color. These do not give off excess carbon monoxide.

Checking Appliances and Other Sources

Gas Stove. First check the pilot light, if present, by looking at it. Is it a nice blue color, or does it have orange or yellow in it? Take a reading above the stove. Is the pilot light giving off excess carbon monoxide? Check the oven for excess CO spillage by taking readings at the rear of the stove top, where the oven vents are located, as well as around the oven door. Place the oven on broil and allow it to warm up for thirty minutes before taking these readings. Turn on the stove burners, check them for color, and take readings over them. A cold pot placed on a gas burner will generate excess CO, as evidenced by the yellow orange flame at the point of contact. As the pot heats up, the flame will reach its intended temperature and glow blue.

Heating Unit and Water Heater. While they are both running, check around the water heater and heating unit for excess CO. Take readings along the flue pipe and exterior of the units. Is the gas flame blue? Are there rust flakes or other debris on top of the gas burner? Either will reduce the flame temperature. If the unit is an oil burner, looking at the flame won't help. It must be checked by a technician.

Recently, manufacturers of high-temperature plastic vent pipes have, in conjunction with the CPSC, announced a recall of their HTPV systems attached to gas or propane furnaces and boilers. These pipes have been found to crack or separate at the joints, leaking CO into the home. It's estimated that 250,000 HTPV systems have been installed in homes around the country. These vent pipes can be identified by their grey or black color. Appearing on stickers or stamped into the vent pipe will be the brand names *Plexvent, Plexvent II,* or *Ultravent*. Only furnace systems in which the vent passes through the house wall are involved, but all such HTPV boiler systems have been recalled. White PVC and CPCV remain unaffected by the recall program. If you encounter HTPV piping during a CO investigation, consider it to be a possible source, and thoroughly test for CO while the heating unit is on. In any case, inform the occupant about the recall and the potential for danger.

Draft Damper. Check the draft hood of the gas heating unit and gas water heater. The presence of soot around its edges indicates the spillage of flue gases.

Check for soot and gas spillage at the draft hood and along the run of the flue pipe.

Perform a draft test by igniting an incense stick or match and holding it near the hood. If the unit is warmed up, and if the smoke from the match or incense stick isn't sucked into the hood, there may be a spillage problem.

Garage. Check for carbon monoxide at the interior door to an attached garage. Take readings at the cracks where the door meets the frame. Even newer cars can give off as much as 30,000 ppm before the catalytic converter warms up. Negative pressure and normal gas diffusion can draw this CO into the home.

Unvented Power Tools and Heaters. Look for any other potential sources of carbon monoxide. Is there an unvented space heater, and has it been in operation? How about gasoline-powered tools or generators? Have they been operated in the home, and were they vented?

Charcoal. Has charcoal been used indoors to cook or heat? Has it been used near an open window or under an awning that might trap the gas and allow it to seep inward? Charcoal gives off substantial amounts of carbon monoxide when it burns.

Fireplace. If there is a fireplace, has it been in use? Smoke stains around the mantel could indicate a downdraft. Is the draft stop open or closed? Is there blockage in the flue? An unused fireplace with an open draft stop might on a windy day allow air to be sucked out of the home, causing another appliance to downdraft as a result.

Chemical Poisoning. Has anyone used chemicals in the vicinity of the detector? Strong chemicals can poison detectors and cause them to sound.

Detector Placement. Don't neglect the detector itself. Improper placement of a detector might cause it to sound unnecessarily. Is it in a high-humidity area, such as outside the bathroom door? Steam might cause it to sound. Is it too close to an appliance? Is it in the kitchen, possibly near the stove? Is it in the garage? Incorrect placement will likely cause the detector to sound for normal discharges of CO that pose no threat to the occupant. Relocating the detector to an appropriate location will prevent a repeat of the alarm.

Taking Appropriate Action

Once you have located the source, your department will have to take appropriate action. This may mean simply to ventilate the building and shut down the malfunctioning appliance. In other cases, the appropriate action may be beyond the means of your department. If the CO were seeping into the home from an underground conduit from a burning manhole, you wouldn't be able to stem the flow. You'd have to call the appropriate utility, as well as extinguish and vent the fire. In New York City, electric utility personnel routinely check adjoining homes for CO at this type of fire, and when they find a problem, they seal the electrical conduits entering the home with an expanding-foam insulation. In the case of an undetectable source, you may have to shut down all potential sources, ventilate, then retest the home.

Duration of the Investigation

How much time can you spend on a CO investigation? Most appliances take ten minutes or more to warm up. An oven may take as much as thirty minutes. This, however, may be time that you don't have. A thorough CO investigation takes about a half-hour to forty-five minutes, but this isn't always practical. Unfortunately, a cursory investigation probably won't lead you to any better conclusion than that the alarm is defective, which it may not be.

If you don't find any CO in the house and lack the time to do a thorough investigation, you should recommend that the occupants have their appliances checked out by the utility or repair service. Suggest that, in the meantime, they leave their windows open. Doing so should provide enough of an air exchange to lessen the severity of a minor CO problem. Before taking such action, consider the possibility of an undiscovered intermittent source of CO in the home. When in doubt, shut off all potential sources, ventilate, and instruct the occupant to have all of the appliances checked.

Evacuating the Premises

When you find high levels of carbon monoxide in the home, you mustn't allow the occupants to remain there. Determine the source, then shut it down. You must ventilate the home. If you cannot locate the source, shutting down all of the fuel-burning appliances and ventilating the home will usually resolve the problem. Once the air clears, close up the home again and retest it for CO. If deemed safe, the occupants may return home. Instruct them not to restart any of the shut appliances until they have been checked by the utility or repair service. Remember to consider a possible exterior source, and always try to reset the detector so that the occupants will have continued protection.

At what point or level of contamination should you evacuate a home? Again, there's no single answer to that question. Different departments use different criteria. Both the New York City and Chicago Fire Departments suggest that occupants leave if the level is between 10 ppm and 99 ppm, and they require evacuation if the level is 100 ppm or above.

SUMMARY

Your investigation isn't complete until you have checked and eliminated all possible sources of CO both inside and outside of the home. Your overall job isn't done until you have eliminated both the source and any residual CO in the home, as well as treated or transported any victims. A carbon monoxide investigation should be methodical, precise, and complete so as not to miss a potentially deadly source. A proper investigation takes time and knowledge, and it must be approached in a professional manner if the resident is to be protected. Always keep in mind that your primary function is to save life. By this criteria, you should treat each CO response as a potentially deadly incident.

STUDY QUESTIONS

1. The vapor density of carbon monoxide is _____.

2. The explosive range of carbon monoxide is _____ to _____.

3. _____ fuels create the most carbon monoxide, _____ fuels generate less, and _____ fuels produce the least.

4. Why will a cold pot temporarily increase the formation of CO when it is placed on the burner of a gas stove?

5. Before reaching operating temperature, a gas oven can give off as much as _____ ppm of carbon monoxide.

6. What is reverse stacking?

7. According to the text, a car that hasn't been warmed up can initially give off _____ ppm or more of carbon monoxide.

8. In terms of percentages, approximately how many U.S. homes are currently equipped with CO detectors?

9. Who should determine exactly how many ambulances are needed when Dispatch receives a call for a CO emergency?

10. True or false: The amount of CO that sets off an alarm would usually, given time, be lethal.

11. To detect low-level sources, it is important that a firefighter use a CO meter capable of sensing gas levels from _____ ppm to at least _____ ppm, and preferably _____ ppm.

12. Which government agency sets the most stringent standards for acceptable levels of CO?

13. When interviewing an occupant at a CO response, you should first take readings _____.

14. If, after a thorough search of the house for CO and defective appliances, you find nothing, you should _____.

15. A carbon monoxide meter that can record CO activity in a home over time and later download the information into a computer is called a _____.

16. A gas flame that isn't properly adjusted and is giving off excess CO will be _____ or _____ in color.

Chapter Twelve
Home CO Detectors

Several different technologies are in use today by the manufacturers of home carbon monoxide detectors. Each detects the gas in a different way and has other unique characteristics. As a responding firefighter, you need to be aware of the different types; their technology; and how, when, and why each sends an alarm. By being familiar with the operation of the various detectors you'll encounter, you'll be better able to determine whether a CO hazard is present. Not being familiar with them will make you more apt to misinterpret an alarm, increasing the chance that you'll leave a dangerous condition uncorrected.

When Dr. Mark Goldstein teamed up with the Quantum group and created the technology that spawned the biomimetic, battery-operated CO detector, it gave the homeowner a reliable and affordable way to be warned of CO in the home. What was this new technology, and how does it sniff out carbon monoxide?

The detector is based on synthetic hemoglobin contained in a small, round disk or gel cell. This gel cell mimics the body's hemoglobin by absorbing ambient carbon monoxide in much the same way that our own hemoglobin absorbs it. As the gel cell absorbs CO, it begins to change color. The color starts out as a translucent orange, but in the presence of CO, it changes to a darker orange and then an olive green. As the amount of CO increases, the color then changes through brown to black. These color changes are monitored by a light-emitting diode, which triggers an alarm once a predetermined level of CO is reached. The more CO that the gel cell absorbs, the darker it becomes. A black coloration indicates that the cell has been killed and must be replaced. Exposure to heavy concentrations of steam may turn the sensor white.

The gel cell, engineered to absorb only carbon monoxide, is similar to the child's toy of geometric blocks and corresponding slots. Only the correct block can enter a specific slot, and the correct block for the gel cell is carbon monoxide. As the gel cell absorbs CO, it also, like the hemoglobin it mimics, releases the gas. Like hemoglobin, it absorbs the gas faster than it releases it, allowing CO to build up on the sensor.

If the sensor is to purge itself of carbon monoxide, you must place it in uncontaminated air. The early model gel cell detector, once it alarmed, typically took from two to twenty-four hours to clear itself of CO in clean air. If it took more than forty-eight hours to clear, the gel had been killed by CO and had to be replaced. The time required to clear the sensor depended on the amount of CO the detector had been exposed to and for how long.

Until the absorbed CO levels dropped below the alarm trigger point, the alarm would continue to sound. To stop the alarm in the early biomimetic detectors, the battery/sensor module had to be removed and the module placed in fresh air until it cleared. Once cleared, it could be placed back into the detector and used again. These old detectors, once they alarmed, left the residents without any protection until the gel cell cleared. The newer models have a reset button that will silence the alarm for six minutes. If, during that six minutes, the ambient carbon monoxide levels drop, the detector will go into its sentry mode and again be ready to warn of a new buildup. If the levels don't drop, the alarm will go off again in about six minutes. The new detectors offer continued protection, even after they go into the alarm mode. In older models, the gel cell is combined with the battery into a single unique model and inserted into a slide-out

This old-type biomimetic detector is shown with its module tray open and the battery/sensor removed. The color chart is used to interpret the color of the gel cells.

A closeup of a new-style battery/sensor module.

drawer in the side of the detector. This unique design prevents the occupant from using the battery for other purposes. As an additional fail safe, the drawer won't close unless the battery module is inserted correctly. This prevents anyone from thinking that the detector is functional after the battery has been removed, a common problem with smoke detectors.

There are two gel cells located on the underside of the module, each in one of two slots. You may need a flashlight to determine their color, though even with a flashlight, the colors aren't easily discerned. Initially, First Alert offered a quick reference card to fire departments to assist them in their investigations. On the card were suggestions for conducting a CO investigation and a color chart used to interpret the tone of the gel. It wasn't a precise indicator, but rather, a rough guide. Although First Alert no longer offers fire departments the color chart, it still offers a card containing useful information for firefighters conducting a CO investigation. You can tell the difference between the old and the new First Alert detectors at a glance. The old ones were white and had the words *carbon monoxide detector* printed in white on their faces. The new ones carry the same legend, but in contrasting print. An additional method of identification is the word *FACOR* printed on the side of the old-type module and the word *NICOR* printed on the new ones. The old module was black and the new ones are white. The color of the module, however, isn't sufficient for determining whether the detector is old or new. The new module will function in the old detector, some stores have sold the old detector with the new module, and some homeowners have replaced old modules with new ones in their old-style detectors. The quickest way to determine which detector you're dealing with is to look for the contrasting lettering. If it is present, you have a new-style detector.

This old-style biomimetic detector has no contrasting lettering, and its battery/sensor module is black.

This newer biomimetic detector has contrasting lettering, and the battery/sensor module is white.

Another difference between the old and new gel cell detector is the type of alarm it emanates. Both the old and the new type sound full alarm with a continuous horn and warn of a low battery level with an intermittent chirp each minute. The new type, however, has an early warning alarm of three to five chirps every five minutes. Knowing which type of alarm the homeowner has and asking him to describe the sound of it can help you determine the reason the alarm sounded and the potential danger to the occupant. If the battery alert sounded, you can advise the occupant to replace the module. If the early warning alarm sounded, you'll know that the level of CO in the house didn't reach dangerous levels, but if the full alarm is sounding, then you may have a dangerous situation. In the latter two cases, the full and warning alarm, CO triggered the alarm and a full investigation is required. In the case of a battery alert, a cursory investigation will suffice.

TYPES OF ALARMS IN GEL-CELL CO DETECTORS

Alarm	Sound
Full alarm*	Continuous
Low battery*	Chirp each minute
Early warning**	3–5 chirps each five minutes.
Replace unit battery/sensor module	2–3 years

*Both old and new type.
**New type only.

NEWER BIOMIMETIC DETECTORS

Portable, Replaceable-Battery CO Alarm. This detector has a rectangular shape with rounded ends. It is powered by a replaceable nine-volt battery and contains a CO sensor that is not replaceable. It's intended either to be wall

Safety-conscious travelers can take this portable biomimetic detector with them on vacation.

mounted or placed on a dresser, table, or countertop. Since CO is only slightly lighter than air and will diffuse throughout a room, the detector can be placed almost anywhere and still function. It's easily carried in a suitcase, and you can expect to hear these detectors alarming in motels as safety-conscious people bring them on vacations and business trips.

PORTABLE BIOMIMETIC ALARM CHARACTERISTICS

Alarm	Sound	Light
Full alarm	Continuous	Flashing red
Low battery	Chirp 2×/min.	Flashing yellow
Replace unit	Chirp 3×/30 sec.	Flashing yellow

Combination Smoke/CO Detector. This new detector has an octagonal shape and contains both an ionization smoke detector and a biomimetic sensor module that has been redesigned to resist low-level alarms. It has a single test/silence button and as a power source uses a single, replaceable nine-volt battery. Both functions of the unit, smoke detector and CO detector, have a distinctive alarm accompanied by an identifying flashing light. The various alarm modes in this detector are different from previous detectors and may initially confuse both the

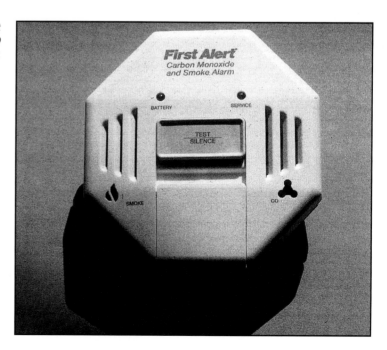

One new addition to the market is a combination smoke/CO detector.

homeowner and the responding firefighter. The two modes, CO and smoke, react differently to the silence button. The smoke alarm is silenced by the button for eight minutes, whereas the CO alarm is silenced for four.

COMBINATION SMOKE/CO ALARM

Alarm	Sound	Light
Smoke	Three loud, consecutive beeps, then a pause	Flashing red flame-shaped light
CO	Single on-off tone, each 1 1/2 seconds long	Flashing red dot pattern
Low battery	Warning chirp	Flashing indicator light
Replace unit	Rapid chirp	Yellow light

SEMICONDUCTOR DETECTORS

There is another affordable technology available that can protect the homeowner from CO. It operates on electronic semiconductor rather than biomimetic technology. The sensor isn't a gel cell; rather, it's a small ceramic component coated with a tin dioxide powder. This powder is formed into a solid by a heating process called sintering. Embedded in the ceramic base are one or two heaters, depending on the manufacturer. This embedded heating element periodically burns off moisture, carbon monoxide, and O_2, as well as other contaminants. The wires coming out of the base are connected to an integrated circuit that monitors the sensor.

Both O_2 and CO are absorbed by the tin dioxide coating. Oxygen increases the electrical resistance of the wire, while carbon monoxide reduces it, allowing electrons to flow more easily. As the concentration of CO increases on the sensor, the lowering of electrical resistance is noted by the microprocessor. Once a predetermined level of CO is reached, the detector will alarm.

The entire cycle, which includes measuring carbon monoxide and burning off moisture and contaminants, takes about two and a half minutes. On average, depending on the manufacturer, the burnoff takes about sixty seconds, and the sampling takes about ninety seconds. Once completed, the cycle repeats, over and over again. If, at any time, the microprocessor detects high levels of CO, the alarm will sound.

Some early semiconductor detectors had reset buttons even before they were required by the revised standard UL 2034. Pressing the reset button caused the burnoff to take place and cleared the sensor of carbon monoxide and other contaminants. The sampling then started again, and if the detector sensed enough CO, it again went into alarm. In fact, this type of alarm, by burning itself off every two and a half minutes, actually resets itself each time it starts a new cycle.

Semiconductor detectors plug directly into an electrical outlet.

Semiconductor detectors use 110-volt house current and can be purchased with a number of options. The detector can either plug directly into an outlet, or it can be wired into the circuitry with electrical cord. It may have a digital readout or it may not. It can be adapted for the hearing impaired by the addition of a flashing light, and it can be purchased with a battery backup. One manufacturer states that its battery backup will last at least eight hours. Another offers a peak-level memory button that will display the highest level of CO registered since the reset button was last pushed. The various features offered vary with the manufacturer.

The typical warranty for such detectors is five years. As these devices age, expect them to become more sensitive, sounding their alarms at lower levels of CO.

This semiconductor detector has a digital display and a built-in electrical cord.

Poisoning a Semiconductor Detector

In February 1996, the International Association of Fire Chiefs issued a warning about semiconductor detectors. The IAFC stated that these detectors might react to high levels of natural gas as if they had been exposed to carbon monoxide. Pat Coughlin, the director of the IAFC's Operation Life Safety, says that a fire department reported that a semiconductor detector reacted to 10,000 ppm of natural gas by alarming as if exposed to carbon monoxide. Natural gas leaks don't normally go unnoticed because of the easily recognized odorant. In this case, however, the odorant had been scrubbed out of the gas as it seeped into the home through an underground pipe break. The danger here is that we might, after checking the home with CO meters, declare the detector defective because we found no CO readings. In this unusual case, we could conceivably leave the homeowner with a potentially dangerous gas leak. This isn't a condition that you will frequently encounter; rather, it's an unusual potential danger of which you must be aware. If your meter says there's no CO present, and if the homeowner's semiconductor is in the alarm mode, check for combustible gas. In this instance, a multigas meter would be handy. Other gases can poison this type of detector, some temporarily and others permanently. The following substances, if present in high concentrations, can temporarily cause a semiconductor detector to falsely sense carbon monoxide: methane, propane, iso-butane, ethylene, ethanol, alcohol, isopropanol, benzene, toluene, ethyl acetate, hydrogen, hydrogen sulfide, sulfur dioxides, aerosol sprays, alcohol-based products, paints, thinners, solvents, adhesives, hair sprays, aftershaves, perfumes, and some cleaning agents.

The following substances can permanently poison a semiconductor detector: sulfur dioxide, hydrogen sulfide, and silicone.

ELECTROCHEMICAL DETECTORS

The electrochemical detector is a relatively new entrant in the home detection market, but the technology on which it is based has been used in commercial gas detection devices for some time. This is the same type of sensor found in the gas detection meters used in the field by the fire service. The current electrochemical detectors haven't been approved by Underwriters Laboratories (UL), but they have been tested by the International Approval Service (IAS).

Features

Sensor. The electrochemical sensor was developed in the 1960s for use in oxygen-sensing devices. The sensor is composed of three electrodes in a pastelike electrolyte and a membrane that is porous only to gas. Carbon monoxide suffuses through this membrane into the electrolyte, reacts with it, and creates a small electrical current. The more CO that is present, the stronger the current will be.

This electrochemical detector operates on a nonreplaceable battery and is easily recognizable by its triangular shape.

Gas Sensitivity. This detector exhibits slight sensitivity to the following gases: hydrogen, ethanol, sulfur dioxide, and hydrogen sulfide. The presence of these gases in the air being sampled may affect the reliability of the unit.

Unique Shape. Some carbon monoxide detectors resemble smoke alarms, especially those manufactured under the old UL standard. (The new standard requires the words *carbon monoxide detector* in contrasting lettering.) The AIM detector's triangular shape is easily recognizable at a glance and won't be mistaken for a smoke detector.

Distinctive Alarm. The audible warning of most early-model CO alarms is similar in sound to that of a smoke detector. This might cause some confusion to the purchaser as well as the responding firefighter. If the occupant of a home has both smoke and CO alarms outside his bedroom door, he won't know which it is when one of them goes off in the middle of the night. AIM's detector goes off with a distinctive baa-beep sound that is easily recognizable. Other manufacturers are now including distinctive-sounding alarms with their newer CO detectors, as well as incorporating the devices in more distinctive shapes.

Warning Alarm. If gas is present but not at dangerous levels, the detector will emit its baa-beep sound every twenty seconds. This will occur before the gas level reaches 100 ppm. Also, its green indicator light, which normally blinks, will shine steadily.

Full Alarm. When the gas level approaches dangerous concentrations, a red light will flash and the alarm will sound with a constant baa-beep signal. An extremely high level of CO will result in a constant alarm and a constant red light. This alarm will sound before the 550 ppm mark. If there is a system error, the alarm will sound and the red light will flash every forty-five seconds.

Data Storage. The AIM detector stores the peak CO level and peak COHb level for the prior twenty-four hours. To retrieve this data, you must have the appropriate AIM gas detection meter, the AIM Safe Air Reader. It isn't, however, a true datalogger, since it doesn't graph the highs and lows of CO exposure against the times that they occurred. A new AIM detector has a digital display that shows the existing CO level, the peak CO level, and the hours that have elapsed since the peak was reached.

Alarm Levels. Although it hasn't been tested by UL, AIM detectors are approved by the American Gas Association. AIM claims its detector will alarm as per the UL 2034 standard at 100, 200, and 400 ppm; that it will give an instantaneous warning when exposed to 550 ppm for two minutes; and that it will continue to alarm until the level drops back below 100 ppm. The detector is also designed to give an early warning of CO buildup.

AIM offers a green home detector that is intended for use by HVAC technicians and fire departments. It's meant to be left overnight in a suspect home, then picked up the next day and analyzed. The green color makes it easily identifiable as fire department property.

Power Source. Neither the battery nor the electrochemical sensor can be replaced. The AIM battery has a five-year life, after which the entire unit must be replaced. Coleman also manufactures an electrochemical detector.

DETECTOR LIFE

Type	Battery Life/Power Source	Life Span
Biomimetic	2–3 yrs., replaceable battery*	5 yrs.
Semiconductor	Plug in	5 yrs.
Electrochemical	5 yrs., nonreplaceable battery** ***	5 yrs.
New technology may increase battery and/or detector life. Refer to individual manual for precise data.		
As semiconductors age, they become more sensitive, so you can expect them to sound more unnecessary alarms.		

*First Alert also manufactures a portable biomimetic detector with the permanent sensor and replaceable nine-volt battery, and a combination smoke/CO detector with a replaceable nine-volt battery.

**Coleman manufactures an electrochemical detector with a replaceable battery and sensor.

***AIM detector.

SUMMARY

The list of detectors, features, and manufacturers is growing. It's important that you stay up to date on the various types of detectors in use so as to best determine why an alarm is sounding. Some knowledge as to how and why these detectors activate, as well as the proper placement for them, will enable you to better advise homeowners. Try to obtain a copy of the instruction booklets for the various types of detectors on the market, since they contain useful information specific to type. Most manufacturers will send you a copy of their manual on request.

STUDY QUESTIONS

1. What sort of detector absorbs ambient CO and is based on synthetic hemoglobin contained in a small gel cell?

2. The more CO that the gel cell absorbs, the _____ it gets.

3. Is the color of the module a sufficient indicator to tell whether a CO detector is of an old or new design?

4. In a semiconductor detector, oxygen _____ the electrical resistance of the tin dioxide sensor, whereas carbon monoxide _____ it.

5. What is the power source used for a semiconductor detector?

6. As semiconductor detectors age, do they become more or less sensitive?

7. Semiconductor detectors may react to high levels of _____ as if they had been exposed to CO.

8. This type of detector is composed of three electrodes in a pastelike electrolyte and a membrane that is porous only to gas.

9. The typical life span of any carbon monoxide detector is about _____ years.

Chapter Thirteen
UL 2034

As of the writing of this book, three generations of carbon monoxide detectors are available and in use. Those built and sold before April 1992 were manufactured to no standard, and many of them used old semiconductor technology. One of them even verbally announced the presence of CO. Underwriters Laboratories began researching the standard for CO detectors in 1989. The resultant standard, UL 2034, was completed and published in April 1992 and revised in October 1995. Detectors manufactured and sold under both the original version as well as the new, revised standard are approved by UL, and you will encounter both in the field. It's important to understand the differences between them. Refer to the owners manuals for specific information about the various brands and models that you'll come across.

CRITERIA OF THE STANDARDS

The 1992 standard allowed both battery and house current to serve as a power supply, and today there are detectors that offer hardwired power supply with battery backup. Both the casing of the detectors and the electronics within must meet the UL standard, and all detectors must meet specific performance criteria, including sensitivity, endurance, audibility, and response to temperature and humidity.

Alarm Level

Underwriters Laboratories specifies when the detectors are expected to sound and when they are to remain still. This is to ensure that a detector won't cry wolf for relatively harmless levels of CO, but that it will respond to potentially dangerous levels of CO before they pose a real threat.

The 1992 version of the standard required that the detectors sound an alarm when the CO present in the air would cause the COHb *in a healthy adult* to reach 10 percent. Note that the criterion for a warning alarm isn't merely based on atmospheric concentrations. At a 10 percent level, a normal, healthy adult should exhibit no symptoms. This means that the onset of the alarm will provide ample time for the occupants to take appropriate action.

For a normal, healthy adult working at a moderate rate, this 10 percent level will be reached in ninety minutes' exposure to 100 ppm of CO. If the CO level is 200 ppm, it will take thirty-five minutes for that same adult to reach 10 percent COHb. At a level of 400 ppm, only fifteen minutes' exposure is required.

Alarm-Level Disclosure

Under the old standard, the detector had to sound before specified levels of CO were reached, but the minimum response level was left undisclosed. The manufacturers of UL-listed CO detectors under the new standard are now required to indicate to the owner both the minimum and maximum levels to which their detectors will respond. These levels vary between the different types and models of detectors available, and this information can be found in each detector's owner's manual.

UL-REQUIRED RESPONSE TIMES

| REQUIRED | | STATED | |
Level UL 2034		Biomimetic	Semiconductor
100 ppm	Within 90 min.	60 min.	Before 90 min.
200 ppm	Within 35 min.	25 min.	Before 30 min.
400 ppm	Within 15 min.	10 min.	Before 15 min.

This table compares UL-required response times with the approximate response time of the biomimetic and semiconductor detectors. The numbers may vary slightly for the various detector brands and models. Check the appropriate owner's manual for specific information on a particular model.

UL 2034 does not require a warning signal, but it does allow manufacturers to include an optional warning signal. If one is present, UL tests it to ensure that it will respond as claimed by the manufacturer.

Resistance to Low-Level Alarms

The original standard requires that the detector activate within the times specified for the specified levels of CO. Thus, a detector would be operating properly if it activated after 89 minutes' exposure to 100 ppm CO. It would also be operating properly if it responded after five minutes. By this standard, an alarm could legitimately sound long before the 10 percent COHb level was reached. So as to reduce unwarranted alarms, UL requires that the detectors *not* sound for a specified period of exposure to specified amounts of CO. This allows levels of the gas to rise to nonlife-threatening levels for short periods of time and then to dissipate without triggering an alarm. This is necessary because various appliances can give off high levels of CO while they are warming up. An alarm sounding in such an instance would be considered unnecessary and a nuisance. The old standard required that the detectors resist an exposure of 15 ppm for eight hours, but because of the high incidence of unwarranted alarms in early models, the standard was revised to a resistance of 15 ppm for thirty days.

Rush-Hour Test

Another source of carbon monoxide that might cause a home detector to sound unnecessarily is a buildup of CO gas resulting from rush-hour traffic. A detector placed in a home near a congested highway might easily sound an alarm during the drive-time buildup. The revised UL standard requires that the detector must pass a rush-hour test. In this test, the detector is exposed to 35 ppm for one hour and then to clean air for six hours. The detector is then again exposed to 35 ppm for one hour and then exposed to clean air for 16 hours. Totaling these hours equals a twenty-four-hour cycle. To pass the test, the detector must not sound an alarm during thirty cycles (thirty days) of the rush-hour test.

Resistance to False Alarms

It's expected that there will be periodic, short-lived high levels of CO in the home, and the UL standard allows for these. Detectors are expected not to sound an alarm when exposed to 100 ppm for sixteen minutes, but they are expected to sound before ninety minutes' exposure to 100 ppm.

SPECIFICATIONS FOR RESISTANCE TO FALSE ALARMS

Concentration in ppm	Exposure in Minutes With No Alarm	
	Old	New
15 +/− 3	15 days	30 days
60 +/− 3	28 min.	28 min.
100 +/− 5	5 min.	16 min.

Alarm Signals

There are three distinct types of alarms, and not all are found on each detector. In all likelihood, an occupant will call the fire department whenever his detector goes off, unaware that his particular device may be capable of sounding several different *types* of alarms. During the interview portion of the investigation, try to determine what type of alarm the occupant heard. If it's still sounding when you arrive, find out its significance. Firefighters must know the various types of alarms, how they sound, and what they indicate.

Trouble Signal. Both the old and the new standard require this alarm, which sounds as a short chirp or beep approximately once a minute. This signal means that a battery needs to be replaced or, in the case of a semiconductor detector, that the unit needs repair or replacement.

Full-Alarm Signal. For most detectors, this is the continuous blaring that indicates CO levels have risen to the danger threshold. In sound, it may be similar to that of a smoke detector, and it's required by both the old and the new standard. The AIM detector sounds full alarm with its distinctive baa-beep sound, and the First Alert combination smoke/CO detector sounds three loud beeps for smoke and an on-off tone for carbon monoxide.

Warning Signal. Some CO detectors issue an early warning signal, indicating that there is a buildup of gas that hasn't yet reached the prescribed activation levels. The warning signal will sound as an intermittent alarm for approximately three to five seconds every three to four minutes. This alarm isn't required for the UL standard, but if present, it must perform as stated by the manufacturer. One manufacturer states that its semiconductor detector will sound in six minutes if it is exposed to 60 ppm and that it isn't tested to alarm below that level. Another manufacturer states that its biomimetic detector will issue an early alarm after thirty minutes' exposure to 100 ppm. For the specifics of a particular model, read the owner's manual. The revised UL 2034 regulations require that the alarm points be stated there.

Reset Button

The original UL standard didn't require the detector to have a reset button, but some semiconductor detectors did anyway. The new standard requires that all detectors have a reset/silence button. If a detector is sounding an alarm and the reset/silence button is pushed, the alarm in most detectors will shut off for six minutes. After the six minutes, it will sound again if the CO level remains at 100 ppm or more. If the CO level has dropped to less than that, the alarm will remain silent.

The original type of biomimetic detector required as much as forty-eight hours to purge itself of carbon monoxide. During that time, the unit would be nonfunctional, and the occupant would receive no warning of a new or continued buildup of carbon monoxide. A reset button affords the homeowner continued protection even after a detector sounds an alarm.

Detector Markings

The original UL standard required that each CO detector be marked with the lettering *carbon monoxide detector*. Unfortunately, on many units, this lettering was the same color as the shell. Glancing quickly, one could miss the letters, and some CO detectors were mistaken for smoke detectors. To correct this problem, UL now mandates that the lettering be in contrasting color to the detector itself.

Instructions to Owners

The original standard required that the owner's manual instruct the purchaser to call the fire department anytime the alarm sounded. The revised standard instructs the owner to follow one course of action if no one exhibits symptoms of CO poisoning and to take another course of action if symptoms are evident. If no symptoms are present, the occupant is instructed to ventilate the interior, then shut down fuel-burning appliances and have them checked by a qualified technician. If symptoms develop, they are instructed to leave the building and call the fire department. The revised standard requires that the occupant call the fire department *only* if symptoms are evident in some of the occupants. This is intended to reduce the workload on the fire service and utility companies by ensuring that they only respond to true emergencies. Unfortunately, instruction manuals are often ignored by consumers. As a result, firefighters may be called anytime an alarm sounds.

COMPARISON OF OLD AND NEW STANDARD

Feature	Original Standard	Revised Standard
Reset button	Not required	Required
Alarm-level disclosure	Not required	Required
Contrasting lettering	Not required	Required
When to call fire dept.	Sounding alarm	Symptoms
Rush-hour test	Not required	Required
Early warning	Not required	Optional, but tested if present

INTERNATIONAL APPROVAL SERVICE

Gas companies, like fire departments, have experienced a significant increase in carbon monoxide calls over the past couple of years. For example, Consumer Power, a Michigan-based utility, responded to 8,400 such calls in 1994 and 13,080 in 1995—a substantial increase for a two-year period. I was told by a local utility official that each CO call cost his company about $150. Multiplying this figure by the number of responses made by Consumer Power in 1995 works out to $1,962,000. Obviously, the gas industry has a vested interest in reducing the number of such calls. In spite of the monetary burden, some gas companies look on the CO issue as an opportunity to provide new service to their customers.

The American Gas Association (AGA) and the Canadian Gas Association (CGA) now test CO detectors under their joint venture, the International Approval Service (IAS). They have developed and published supplemental standards for these detectors. Their intention is to reduce unnecessary low-level alarms by having detectors comply with both the IAS and the UL standards while still protecting consumers from dangerous levels of CO. (Note: There is conflicting opinion as to whether low-level exposure to CO is harmful and as to what constitutes an unnecessary alarm.) The standard, IAS US Requirement Number 6-96, was issued October 11, 1996. This organization is affiliated with the gas industry, not Underwriters Laboratories. The IAS standard, as well as recommendations from the Consumer Product Safety Commission, has been submitted to UL for their consideration. At the time of this writing, no detector has been approved by both UL and IAS.

IAS Alarm Levels

The IAS requires that an alarm activate in response to an exposure time versus a concentration between 5 percent COHb and 10 percent COHb. It further states that there shall be no early warning signal. It allows an alarm only in response to the stipulated levels of carbon monoxide.

IAS REQUIREMENTS
CO CONCENTRATION VS. TIME

Concentrations in ppm	Time to Alarm in Minutes (10% COHb)	Minimum Response in Minutes (5% COHb)
65 +/− 3	245	65
100 +/− 5	90	35
200 +/− 5	35	16
300 +/− 5	20	9
400 +/− 5	15	7
600 +/− 5	9	

False-Alarm Resistance

The IAS standard requires that the alarm not sound before the occupants' COHb levels reach 5 percent. In addition, the detector should resist sounding an alarm according to the following table.

IAS NO-ALARM REQUIREMENTS

Concentration in ppm	Exposure Time (No Alarm)
30 +/– 3	30 days
50 +/– 3	See explanation below*
100 +/– 3	35 minutes

* The detector should resist an alarm after an exposure of 50 ppm for one hour, followed by six hours in clean air, followed by 50 ppm for one hour, followed by 16 hours in clean air.

Instructions

The IAS standard instructs the consumer, in the event of an alarm, to evacuate the building and to call the fire department. Consumers are further instructed not to reenter the building without the permission of the responding emergency services personnel.

CONSUMER PRODUCT SAFETY COMMISSION

The Consumer Product Safety Commission recommends the elimination of an early warning signal because it considers the signal confusing to the purchaser. It further recommends that the detectors alarm at the following levels:

CPSC SUGGESTED ALARM ACTIVATION LEVELS

70 ppm	240 min.
100 ppm	90 min.
200 ppm	35 min.
400 ppm	15 min.

CPSC SUGGESTED FALSE-ALARM RESISTANCE POINTS

30 +/– 3 ppm	30 days
50 +/– 3 ppm	1 hour

The CPSC documents that ambient levels of carbon monoxide can be high in certain areas of the country. The agency points out that ambient levels, when combined with the normal amount of CO given off by unvented gas appliances in the home, can cause home detectors to sound even though the gas level isn't an immediate safety hazard. In Chicago in 1994, during a thermal inversion, CO detectors throughout the city responded to the 10 ppm of carbon monoxide that was trapped with other pollutants, causing the Chicago Fire Department to respond to more than 1,800 alarms in a twenty-four-hour period. The majority of those alarms were received during the first twelve hours or so. Thermal inversions aren't unique to Chicago, nor are the phenomena that they bring. The changes suggested by the CPSC would lessen the likelihood that poor air quality causes alarms to sound.

Since certain risk groups require more stringent protection, the CPSC further suggests that detectors be developed and sold specifically for their use.

The organization tests and rates appliances, and as such has an interest in the carbon monoxide emissions from gas-burning ones. The CPSC believes that, currently, a properly functioning unvented gas appliance can trigger a CO alarm—i.e., that the existing false-alarm resistance points are unrealistically low and that the levels set by UL translate into numerous unnecessary alarms.

The CPSC also recommends that the list of gases that CO detectors should resist be expanded to include the following: amines (ammonia), aromatic hydrocarbons (xylene and toluene), and halogenated hydrocarbons (methylene chloride, trichlorethane, and partially oxidized hydrocarbons, including acetone and ethanol). Because a properly operating gas appliance produces carbon dioxide, the CPSC further suggests that detectors should resist 5,000 ppm of CO_2 rather than the 1,000 ppm as currently specified.

Those detectors that feature a digital readout of ambient carbon monoxide start displaying CO at varying levels and with different accuracy. One detector might not display the level of CO until it reaches 50 ppm, while another might show a reading of 5 ppm. The CPSC recommends that a minimum standard of accuracy for digital readouts be established. The decision makers at Underwriters Laboratories are considering the recommendations of the AGA and CPSC as they consider changing their existing standard. Whether they will implement the suggested changes or come up with new ones isn't clear. What is clear is that, as existing detectors are reevaluated and as new technologies are introduced, the standard will continue to evolve. It's important that you and your department stay on top of both the standards and the technologies so as to respond properly to CO calls.

PUBLIC EDUCATION

There is much that can be done to reduce the number of responses resulting from the proliferation of CO detectors. As the public is made aware of the dangers of carbon monoxide by the media and public information messages, we should also be explaining to them the proper placement and maintenance of such devices.

A woman and her husband in Ohio, after hearing about the dangers of carbon monoxide, decided to buy one of the first battery-operated detectors to protect themselves and their five children. Sometime after making the purchase, and while the husband was away on business, the detector went off in the early morning. One of the children woke up first and told her mother about it. Seeing and smelling nothing, the mother took the battery out of the detector and went back to sleep. "It's broken," she thought. The next day, she tried to replace the battery, but the alarm sounded again, so she pulled the battery a second time. This time, she left it out. Over the course of the next few days, she and her children began to feel sick. Complaining of nausea and headaches, they went to the doctor. He told them that they had the flu and gave them a prescription for antibiotics. They took their medicine but continued to feel sick. The husband, away on business most of the time, didn't share these symptoms. One day, they noticed a strange smell in the house and called the gas company. After testing the air, the serviceman shut down the gas burner and called the fire department. The fire department checked for CO and found high levels in the home. The firefighters asked whether anyone felt ill and found that all of the occupants were feeling symptoms of CO poisoning. The problem was a blocked flue that had caused the fumes to spill into the home. Because the adults in the household didn't understand the nature and dangers of carbon monoxide, nor what to do when the alarm sounded, the family was made ill and put in danger of death. Once the problem was corrected, all of the occupants made a miraculous recovery and had no recurrence of symptoms.

Sometime after this incident, the same family had to move in with a relative while work was being done in their own home. All of their belongings were packed up in boxes. Shortly after moving in, their CO detector (the same one that had alerted them to the previous danger, and now packed in a box in the basement) sounded a warning alarm. This time, educated to the dangers, the woman immediately called the fire department. Again, the responders found dangerous levels of carbon monoxide.

Both times, the detector did what it was designed to do. Luckily, the family survived the first episode and learned the lesson.

We try to educate the public about the need for smoke detectors, the dangers of fire, and what to do when a fire occurs in the home. This education has helped cut down on fire deaths. We should offer the public the same type of education with respect to carbon monoxide and what the homeowner can do to protect himself from it.

UL-Approved Detectors

Urge the public to purchase UL-approved detectors. There are gas-absorbing cards being sold as CO detectors. The card contains a small dot that changes color as it absorbs carbon monoxide, thereby warning that the gas is present. I have found advertisements for such cards in child safety catalogs alongside UL-approved detectors. An unsuspecting parent might purchase such a card thinking that it will offer, for less money, the same level of protection. It will not. A consumer must look at it to note the change in color, and it won't wake you up if the CO buildup occurs while you are asleep and most vulnerable. A UL-approved detector assures the purchaser of an acceptable level of reliability and will sound an audible alarm when the ambient CO reaches a determined level.

Public Service Announcements

Prior to the onset of winter and during winter storms, warnings describing the dangers of carbon monoxide should be broadcast on radio and television. During the winter, people seal themselves in their homes, along with the carbon monoxide being produced by their appliances. The power failures that can accompany winter storms often result in a loss of heat. As a result, homeowners leave gas stoves and space heaters on for inordinate periods of time in an effort to stay warm. Two different studies have indicated that as many as fifty percent of all urban low-income dwellings are heated with their ranges. Undoubtedly, some nonurban, higher-income homeowners do the same when coping with power outages. I have frequently encountered apartments being heated by gas ovens because landlords haven't provided adequate heat or because a winter storm has damaged power lines. This presents a large population that is at risk of CO poisoning from the improper use of their gas appliances.

If a segment of your local population is non-English-speaking, you'll have to ensure that the warnings are printed and broadcast appropriately.

A good time to pass along information about carbon monoxide is when it's in the news. If someone in your community dies as a result of CO poisoning, the local stores probably won't be able to keep CO detectors in stock. A high-profile death, such as that of Vitis Gerulitis, also presents a good time for public education. Such incidents ordinarily make people wonder whether they are safe from the threat. Consequently, they'll be more receptive to any information that you give them. Remember, the public has a short attention span, so you must be ready with your public service announcement before the news-making incident occurs.

Detector Placement. As the heating season approaches, make information available to the public as to the proper placement of CO detectors. A carbon monoxide detector should be placed similarly to a smoke detector. Don't place one in the dead air space that exists within six inches of where a wall meets the ceiling. Don't place it behind curtains or furniture or anything else that would block the airflow to it, nor should you locate it near an HVAC supply register or near a fan. Depending on the make and model, don't place it within five to twenty feet of an appliance. I favor the twenty-foot figure, since it will help cut down on unnecessary alarms. It's ill advised to place a CO detector in the kitchen or anywhere else prone to grease, dust, and chemicals. Also avoid exposing it to cleaning fluids or paint thinner. Don't place it in an uninsulated space or on an uninsulated wall. Like a smoke detector, the CO detector is meant to give an audible warning to occupants in the event of a life-threatening danger. If a sleeping occupant cannot hear the alarm, then the alarm is useless. The CPSC recommends that a CO detector be placed on each level of a home near the sleeping area. One that's placed in the basement may not be heard by the sleeping occupant one or two floors above. Manufacturers recommend that the detector be placed wherever the occupants of the home congregate. A good time to relay all such information is when you have responded to an alarm that has sounded as a result of improper placement. This is your opportunity to point out the better places to mount a detector, hopefully to avoid responding again for another unnecessary alarm.

As of this writing, the NFPA is working on a standard for CO detectors. This new standard, NFPA 720, will recommend the placement of CO detectors within the home.

Homeowner Response to an Alarm. Explain the appropriate action to take when a CO detector goes off. Not every alarm indicates an emergency that requires a fire department response. As stated above, the manufacturers' instructions on the newest alarms state that the only time the fire department need be called is when symptoms are felt by someone in the home. If symptoms are felt or suspected, the occupant should vacate the premises immediately and call the local fire department. Depending on the number of calls that your department receives and the type of dwellings in your area, you may or may not want to reinforce these instructions with public service announcements. What might be good advice to a homeowner might not be the best to give to an apartment dweller or someone who lives in an attached row of buildings. The carbon monoxide might be seeping in from another apartment or building. An apartment dweller often has no control of the furnace, and as a result, cannot shut off that source. Although ventilating the apartment or attached home will temporarily clear the carbon monoxide from the air, it will build up again once the windows have been closed. In a multifamily building, not calling the fire department for a sounding alarm might be dangerous. If the occupants with the alarm vent their windows, they may be fine and their alarm may no longer sound, but what about those in adjoining apartments or next door who don't have a CO detector and who may be suffering from the effects of the gas? It may be wiser to call the fire department and let members conduct an investigation as to the cause and extent of the alarm. Several years ago on Long Island, a faulty furnace sent twelve people to the hospital. The furnace serviced a complex of buildings, and as a result, the escaping CO threatened many families. In such a situation, instructing residents not to call the fire department, but instead to silence their detector and vent their individual apartment, could prove deadly to many other residents. There is an obvious need here to call the fire department and have them make an examination.

One problem in explaining to the public the proper action to take is that the manufacturers, government agencies, and medical authorities don't agree on what the proper action is. The instructions in owner's manuals have changed, and in all likelihood, they'll change again. Any information that we convey to the public must be as accurate as we know it to be today, and we must consider it in terms of possible future liability.

Annual Maintenance. Explain the benefits of annual servicing of fuel-burning appliances and flues. If you encourage homeowners to get their appliances checked and serviced before the heating season, it's less likely that you'll have to respond to them for a malfunction. A flue that has been checked and cleaned is less likely to become blocked and spill gas into the home, and a properly maintained appliance will give off less CO than a neglected one. Routine cleaning and maintenance will go a long way toward protecting people and reducing the number of nonemergency calls they make.

Collaboration

Your department may want to consider teaming up with other government agencies or private entities, such as the utility or fuel oil companies, to get the message out to the public. The more ways that you can convey information, the greater the likelihood that your message will be heard.

KEEP UP TO DATE

By the time you read this book, some of the material on carbon monoxide detectors and standards may have already changed. New technology and brand names may have come on the market. Fortunately, the manufacturers of detectors will help you become aware as they advertise their products. When you hear of a new product, contact the manufacturer and ask for a sample, or at least an owner's manual. There may be new types of appliances and, as a result, new sources of CO poisoning. If you encounter a new CO source or detector, you should make its presence known to other firefighters, as well as other departments. Only by keeping ourselves up to date can we remain professionals to those we serve.

SUMMARY

From the start, there was confusion over the issue of carbon monoxide responses. How did the detectors work? What was a dangerous level of carbon monoxide? Some of the early detectors went off at such low levels that they were thought to be malfunctioning. As a result, some firefighters came to believe that all CO detectors are unreliable, yet those who hold such an opinion need to rethink their position. The new standard changed the minimum alarm thresholds and has eliminated much of the problem of low-level alarms. The technology of the detectors has improved, and new manufacturers have come to the marketplace. Responding firefighters may be faced with a array of old and new technologies, as well as a myriad of brand names, but knowing whether a given unit was produced under the old UL 2034 standard or the new one is an important step toward determining why its alarm has gone off.

When a homeowner calls the fire department because his CO detector is making noise, he probably won't know under which standard the unit was produced. He won't know whether it has a warning alarm or not. He probably never read the instructions that came with it, and he may likely have placed the device in an improper location. Arriving firefighters will be expected to be knowledgeable about all of these aspects, to take the appropriate action, and to offer solid advice. If you cannot accurately provide that advice, you may be placing the occupant in danger, or your offhanded guesswork regarding detector placement might result in an abundance of unwarranted calls to that same location.

The role of the modern firefighter isn't only to put out fires, it's also to educate. Just as we have successfully convinced a majority of homeowners to use smoke detectors, we must also convince them to use CO detectors. By fully understanding the subject matter, we can better serve the public, reducing their risk of illness or death from carbon monoxide, while lowering the number of unnecessary responses that we make.

STUDY QUESTIONS

1. Originally published in April 1992, UL 2034 was revised in October of what year?

2. The 1992 version of the standard required that the detectors sound an alarm when the CO present in the air would cause the COHb in a healthy adult to reach _____ percent.

3. At 400 ppm, UL 2034 requires a CO detector to sound an alarm within what length of time?

4. True or false: UL requires that detectors remain silent for a specified period of exposure to specified amounts of CO.

5. The revised UL standard requires that a detector pass the rush-hour test for how many days in a row?

6. Does the revised standard require all CO detectors to have a reset button?

7. The International Approval Service (IAS) standard requires that an alarm not sound before the occupants' COHb levels reach _____ percent.

8. True or false: Carbon monoxide detectors may be triggered by a thermal inversion.

Answers to Study Questions

Chapter One

1. Polyvinyl chloride (PVC).
2. Ballast.
3. PCB.
4. False.
5. 1982.
6. The belts of major appliances.
7. Vent duct.
8. True.
9. Thermal imaging camera.
10. No. Curtail the power by using the main shutoff instead.
11. No. It's possible that power is being pirated into the building, bypassing both the meter and the main shutoff.
12. True.
13. Yes.
14. (1) Explosion, (2) PCB contamination, (3) downed wires, (4) electrocution.
15. Live.
16. No. Even a seemingly dry rope or pike pole can contain contaminants that might conduct electricity.
17. One span (the distance between two poles).
18. Tie off the open butt of a hoseline near the opening, then withdraw all personnel to a safe distance before turning on the water.
19. Salt.
20. Flammable and explosive.
21. Carbon monoxide.
22. Appropriately trained utility personnel.

Chapter Two

1. Oil and natural gas.
2. Six.
3. 275 gallons.
4. Atomized.
5. Ninety.
6. Delayed ignition (puffback).
7. Afterfire.
8. Pulsation.
9. Gas leak.
10. No. Closing the furnace door can cause severe pulsation.
11. Let it burn out on its own.
12. A blocked or disconnected flue or chimney.

13. Make the occupants of the building sick.
14. (1) 210°F, (2) 150 psi.
15. Steam or hot water discharging from the temperature-pressure valve, or steam rather than water discharging from an opened hot water faucet.
16. A red-hot water heater and the smell of hot metal.
17. Delayed ignition.

Chapter Three

1. (1) Methane, (2) ethane.
2. Asphyxiation.
3. Mercaptan.
4. Two-thirds.
5. 350 psi.
6. 1/4 psi.
7. Causes flameout.
8. Doubles.
9. New.
10. Digital flammable gas detector.
11. (1) The gas flame can either be blown out, possibly resulting in a gas leak, or (2) the flame may grow, possibly to dangerous heights.
12. Delayed ignition.
13. Shut off the escaping gas.
14. Yes.
15. Explosion and structural collapse.
16. True.
17. No, because the gas has probably accumulated near the ceiling.
18. Meter quarter-turn wingcock.
19. False. Plastic pipe, specifically because it is nonconductive, will allow a static charge to build up as gas flows freely through a break in the pipe. No static charge will build up on a conductive metal pipe.
20. Yes, but you must consider static electricity as a possible source of ignition and take proper precautions.

Chapter Four

1. 62.4 lbs.
2. 30' × 45' = 1,350 sq. ft. × 0.25' (depth of water) = 337.5 cubic ft. × 62.4 lbs. (weight of a cubic foot of water) = 21,060 lbs.
3. Resist kinking.
4. Pumper.
5. If the water reaches an electrical outlet, it can cause a short that trips the circuit breaker, or it can start a fire. If the breaker doesn't trip, the area near the submerged outlet can become electrically charged.
6. False. Water that is still in the ceiling, in walls, and on wires and appliances can cause switches, appliances, and even countertops and floors to become electrically charged.
7. True, since the handle of a pike pole is less likely than the hook-shaped head to cause the ceiling to collapse as you withdraw it.
8. Yes. Shutting a hydrant too quickly can create a water hammer, which may burst the water main.
9. Verify that the leak is from the main.
10. The possibility exists that the road has been undermined and that the weight of the apparatus will collapse it.

Chapter Five

1. Haz mat incident.
2. From the sides.
3. Hydrogen and oxygen.

4. 4 percent lower to 71 percent upper.
5. Diesel.
6. The fuel line, which may convey gasoline at 30 psi, will retain its pressure even after the engine has been shut off.
7. Both power steering fluid and ATF are flammable when released under pressure in the form of spray. In a modern vehicle, power steering fluid can be pressurized to as much as 200 psi.
8. A BLEVE.
9. No. A CNG tank can't experience a BLEVE, since it contains a gas, not a liquid.
10. No. There may be a pedestrian pinned underneath the vehicle.
11. When the car is found in a remote area.
12. Uphill.
13. (1) From the passenger compartment, if the backseat has been burned out or removed, and (2) by breaking a taillight and inserting a nozzle into the hole.
14. Structure fire.
15. By clamping it onto the track below the door's rollers.
16. To beyond the estimated stopping range of oncoming traffic, and slanting to cover the farthest lane occupied by emergency personnel.

Chapter Six

1. The type of food being cooked.
2. False. The method of forcible entry should be decided on a case-by-case basis.
3. Yes.
4. Nausea, stomach pains, vomiting, oxygen deficiency in the blood, altered blood pressure, headaches, delayed onset of symptoms.
5. Two-thirds at the top and one-third at the bottom.
6. It is no longer a food-on-the-stove call, but a structure fire.
7. Yes. There may have been open flame before you arrived.
8. False. The hot oil can vaporize, ignite, and flare up even higher.
9. Yes. The steam that's generated after you close the door will probably smother the fire.
10. Shut off the gas at the meter.

Chapter Seven

1. The entire apartment, the adjoining apartments, and the apartment above, at minimum.
2. Thoroughly soak the mattress with a handline.
3. Cut open the mattress to extinguish fire at the core.
4. False. Foam mattresses and those with foam pads can continue to generate combustible gases even after the active flames have been extinguished.
5. Fold the mattress in half so that the burned side is toward the fold, then tie it closed.
6. False.

Chapter Eight

1. Use the reach of your line and attack the fire with the wind at your back.
2. Any dumpster is prone to the illegal dumping of hazardous materials.

3. False.
4. The firefighter.
5. Heavy equipment.
6. Shine bright lights up from below.
7. By monitoring the runoff.
8. To warn firefighters at future incidents of the building's relative stability at the time it was marked.

Chapter Nine

1. 200 times.
2. False.
3. UL 2034.
4. (1) Improved energy efficiency has made modern homes more airtight, and (2) many CO-related deaths and illnesses in previous years weren't attributed to carbon monoxide.
5. About 1,500 deaths; more than 10,000 illnesses.

Chapter Ten

1. Four.
2. Breathing rate.
3. The ambient level of CO.
4. 20 percent.
5. Greater and faster.
6. Those that require the most oxygen—e.g., the cardiovascular system and the brain.
7. No.
8. The flu.
9. Eighty minutes.
10. True.

Chapter Eleven

1. 0.96.
2. 12.5 to 74 percent.
3. (1) Solid, (2) liquid, (3) gas.

4. Because it lowers the temperature of the flame.
5. 800 ppm.
6. A mechanically induced downdraft.
7. 30,000 ppm.
8. Six percent.
9. The officer of the first-arriving fire company.
10. False.
11. (1) 0 ppm, (2) 200 ppm, (3) 999 ppm.
12. The EPA.
13. At the door.
14. Set up a worst-case scenario.
15. Datalogger.
16. Yellow or orange.

Chapter Twelve

1. Biomimetic.
2. Darker.
3. No.
4. (1) Increases, (2) decreases.
5. 110-volt house current.
6. More sensitive.
7. Natural gas.
8. Electrochemical.
9. Five years.

Chapter Thirteen

1. 1995.
2. 10 percent.
3. Fifteen minutes.
4. True.
5. Thirty.
6. Yes.
7. 5 percent.
8. True.

Index

A

Afterfire, *51–2*
AIM, *218–9, 226*
Air slug, *70*
Alternative fuels (automotive), *110–1*
Aluminum siding, *20*
American Gas Association (AGA), *170, 219, 228, 230*
Appliance quarter-turn shutoff, *81*
Aquastat, *48, 49*
Atomization, *50*

B

Ballast, *6–8*
Barometric draft damper, *47*
Batteries (automotive), *107–8, 115*
Blast mat, *38*
BLEVE, *110, 111*
Braking distance, *122*
Brooklyn Union Gas Company, *79*
Bumpers, *107*
Butane, *68*

C

Canadian Gas Association (CGA), *170, 228*
Capacitor, *6, 7*
Carbon monoxide, *43, 60, 67, 110, 167–235*
    and appliances, *171, 183–4, 199, 200, 202, 204*
    common sources of, *183–9*
    and energy efficiency, *171, 188–9*
    explosivity of, *183*
    and furnaces, *187–8*
    long-term effects of, *131, 176*
    and manufactured gas, *67, 84*
    poisoning, *43, 60, 175–81, 198–9*
    response protocols, *199–207*
    safe levels of, *197–8*
    signs of, *203–4*
    symptoms of, *178–80, 198*
    treatment for, *180–1*
    and vehicle fires, *110*
    and vehicles, *188, 189, 205*
Carbon monoxide detectors, *167–72, 189–90, 206, 209–34*
    biomimetic, *209–215, 219, 224, 226*
    card type, *231–2*
    combination, *214–5*
    electrochemical, *217–9*
    semiconductor, *215–7, 219, 224, 226*
    placement of, *232–3*
Carbon monoxide meter, *37, 169, 192–5, 199–200*
Charcoal, *189, 205*

Chicago Fire Department, *197, 207*
Combustible gas indicator, *79*
Combustion explosion, *71–2, 75, 77, 86*
Compressed natural gas, *110, 111*
Consolidated Edison, *40*
Consumer Power, *227–8*
Consumer Product Safety Commission (CPSC), *43, 81, 170, 197, 204, 228, 229–30, 233*
Creosote, *150*
Curb valve, *82–3, 104*

D

Danger zone
    electrical, *27–8*
    manhole incidents, *35, 37*
    natural gas, *89–90*
    water main incidents, *105*
Datalogger, *194, 202–3, 219*
Delayed ignition (puffback)
    gas water heater, *64*
    natural gas, *74–5*
    oil heat, *49–51, 54, 55, 56–7, 59, 60, 185*
Dewatering operations, *95–6, 98*
Dielectric fluid, *7*
Dishwashers, *10*
Downdraft, *64, 185–7, 200*
Draft damper, *204–5*
Drip pot, *86–7*
Drive belt, *9, 10–1*
Dry-chemical extinguisher, *10, 51, 55, 57–8, 111, 115, 121*
Dryers, *10, 11–3, 81, 187, 202*
Dumpsters, *152–4*

E

Eductor, *98, 100*
Electrical emergencies, *5–40*
    appliances, *9–13*
    downed wires, *20–1, 23–8*
    fixtures, *13–4*
    manholes, *28–40*
    meter and, *19–20*
    odors, *5–10*
    on-pole, *22–3*

Electrical emergencies, *continued*
    outlets, *13–4*
    utility service, *17–20*
    vehicles and, *23–5*
    weather and, *25, 32*
Electrical fires
    causes of, *14–5*
    extension of, *15–7*
Electric fans, *186, 187, 202*
Emergency shutoff switch (oil burner), *48, 55, 57*
Environmental Protection Agency (EPA), *170, 176, 197*
Ethane, *67*

F

Fire Department Instructors Conference (FDIC), *169*
Fire Department of New York, *86–7, 197, 207*
Fireplaces, *179, 183, 203, 205–6*
First Alert, *168, 178–9, 211, 226*
Flame-retention burner, *52*
Flares, *122–3*
Fluorescent lights, *6–8*
Flushometer, *97*
Forcible entry
    of automobile trunks, *113*
    of dwellings, *78, 128–30*
    of unoccupied buildings, *162*
Fuel injection (automotive), *108*
Fuel oil, *44*
Fuel tank (automotive), *108, 154*

G

Garage doors, *119–20*
Gas detectors (natural gas), *72–3, 79–80, 89, 169*
Gas oven, *184, 199*
Gerulitis, Vitis, *192, 232*
Goldstein, Dr. Mark, *171–2, 209*

H

Halogen lamps, *9*
Hard starting (oil), *50–1*

Heat exchanger, *47*
Heavy equipment, *157–9*
High-limit switch, *8, 9*
HTPV systems, *204*
Hydraulic fluid, *109*
Hydrogen, *108, 217*
Hyperbaric oxygen, *177, 180–1*

I

Ignition system (oil), *48*
International Approval Service (IAS), *170, 217, 227–9*
International Association of Fire Chiefs (IAFC), *168–9, 201, 217*

J

Junkyards, *154–9*

K

Kitchen fires, *127–38*
    extension of, *133–7*
    gaining entry, *127–30, 135*
    microwaves, *138*
    mitigation of, *132–3, 135–6*
    overhaul, *136–7*
    safety issues, *130*
    search, *130, 135*
    stove location, *132*
    ventilation of, *131–2, 136*

L

Limit controls (oil), *48*
Lint, *11–3*
Liquid petroleum gas, *67, 68*
Low-water cutoff
    oil burner, *48–9*
    water heater, *63*

M

Magnesium, *121*
Manholes (electrical), *28–40*
    corrosion and, *33*

Manholes (electrical), *continued*
    covers, *34–5, 37*
    fumes from, *33–4*
    response to, *35–40*
    types of, *30*
Manufactured gas, *67, 84*
Maryland Institute for Emergency Medical Services, *176*
Mattress fires, *141–6*
    mattress removal, *144–6*
    and overhaul, *142–3*
    and search, *142*
Mercaptan, *67, 72, 89*
Meter
    electrical, *19, 41, 85*
    gas, *64, 70, 74, 77, 79, 80, 81–2, 89, 137*
Meter quarter-turn wingcock, *82*
Methane, *67, 68, 217*
Methylglycol, *109*
Microwave ovens, *138*
Multigas meter, *193–4*

N

Natural gas
    asphyxiation by, *67, 84–5*
    composition of, *67*
    distribution systems, *68–70*
    fires, *75-7*
    hazards of, *70-5*
    indoor leaks, *77–86*
    odorization of, *67–8, 72–3*
    outdoor leaks, *86–90*
    pressurization of, *69–70*
    and semiconductor CO detectors, *217*
    and static hazard, *87–8*
    underground migration of, *87, 88–9*
National Fire Protection Association (NFPA), *43, 233*
New York City Fire Department Institute, *169*
NFPA 720, *233*
Nitrogen dioxide, *131*
North American Congress of Clinical Toxicology, *176*
Nozzle assembly (oil system), *47*

O

Occupational Safety and Health Administration (OSHA), *170, 197*
Oil burner emergencies, *49–57*
Oil heating system, *44–61*
    components of, *44–9*
    emergency response, *54–9*
    fires, *57–9*
    problems with, *49–54, 60*
    spills, *60-1*
Operation Life Safety, *168, 217*

P

Parapets, *94*
PCB (polychlorinated biphenyl), *6, 7, 22–3, 30, 32, 152*
Peak shaving, *67, 68*
Perception time, *122*
Plastic gas pipe, *87–8*
Pressuretrol, *49*
Primary control (oil), *49*
Primary search, *111, 130*
Propane, *43, 44, 64, 169, 183, 204, 217*
    and automobiles, *110-1*
    and natural gas, *67, 68, 79*
Public service announcements, *232*
Puffback, see Delayed ignition
Pull box, *17*
Pulsation, *52, 54*
PVC (polyvinyl chloride), *6, 150, 204*

Q

Quantum Group, *172, 209*

R

Radiators (automotive), *109–10*
Rapid intervention team, *160*
Reaction time, *122*
Recessed lighting, *8*
Refrigerators, *10, 183*
Regulator (natural gas), *64, 70, 71, 73–4*
Reverse stacking, *186, 187, 202*
Roof drains, *94–6*
Rotary cup burner, *50*
Rush-hour test, *225*

S

Safety shutoff (water heater), *63*
Salvage pump, *96, 98, 100*
Secondary search, *111, 130, 142, 143*
Single-gas meter, *193–4*
Siphoning, *95–6*
Smoke ejector, *130*
Space heaters, *184, 205, 232*
Stack switch, *49, 50*
Stopping distance, *122*
Stove (gas), *81, 132, 133, 137, 184, 199, 200, 204, 232*
Street collapse, *102–4*
Street valve, *83*
Stuffed furniture, *146*
Sulfur, *131*
Sulfuric acid, *107–8*

T

Temperature-pressure valve, *61–3*
Thermal cutoff switch, *9*
Thermal imaging camera, *13, 16–7*
Thermal protection, *8*
Thermostat, *48*
Tires, *108*
Transformers
    appliance, *9*
    fluorescent ballast, *6–8*
    oil heating system, *47*
    pole-mounted, *22–3, 24, 25*
    underground, *30-2, 38–9*
Trash fires
    apparatus placement, *151*
    in dumpsters, *152–4*
    and exposures, *154, 161–2*
    in junkyards, *154–9*
    life hazards of, *151, 154, 156*
    smoke hazards of, *150-1*
    structural, *159–61*

U

UL 2034, *170, 215, 219, 223–4*
Underwriters Laboratories (UL), *8, 170, 219, 223-34*

V

Vacant buildings, *159–61*
Vehicle fires, *107–24*
    and alternative fuels, *110-1*
    apparatus placement, *114, 123–4*
    derelict vehicles, *117*
    and garages, *117–21*
    hazards of, *107–10*
    on highways, *121–4*
    overhaul of, *115, 116*
    search of, *111–3, 115*
Ventilation (of dwellings), *85–6, 121, 130, 131–2, 136, 233*
Vent pipe
    gas system, *73–4*
    oil system, *44–5, 46*
Visual flame detector, *49, 50*

W

Wall oven, *136*
Washing machines, *10-1*
Water emergencies, *93–105*
    and ceilings, *94, 101*
    and electricity, *98, 99, 100*
    and elevators, *101*
    flooded basements, *98–100*
    flooded roofs, *94–6*
    and heating systems, *99*
    life hazards of, *98–9*
    plumbing, *96–7, 100*
    safety concerns, *100-1*
    water mains, *102–5*
Water hammer, *102*
Water heaters, *61–4, 204–5*
Water, weight of, *95*
White ghost, *52–4, 58, 59*